Introduction to Quantum Computation

Ioan Burda

Introduction to Quantum Computation

Copyright © 2005 Ioan Burda
All rights reserved.

Universal Publishers
Boca Raton, Florida • USA
2005

ISBN: 1-58112- 466-X

www.universal-publishers.com

For my parents and wife

Contents

1 Introduction

The theory of computation has been long considered a completely theoretical field, detached from physics. One of the people most directly responsible for this concept of computing machines is Alan Turing (1912-1954). Turing and others early giants proposed mathematical models for computing which allowed for the study of algorithms and in absence of any particular computer hardware. This abstraction has proved invaluable in the field of computer science. Turing's model (Turing 1936) is called a Turing machine.

When examining what sort of problems can be solved by a computer, one need only to examine a Turing machine, and not one of the millions of potential computing devices to determine if a computation is possible. If a Turing machine can perform the computation, then it is computable. If a Turing machine can not, then the function can not be computed by any classical computer. Church's thesis (Church 1937) states:

Any physical computing device can be simulated by a Turing machine in a number of steps polynomial in the resources used by the computing device.

This thesis gives us insight into the *power* of computing machines. If a computer theoretically can solve all the problems a Turing machine can solve (given enough memory and time) then it is as powerful as a Turing machine.

An algorithm can be characterized by the number of operations and amount of memory it requires to compute an answer given an input of size n. These characterizations of the algorithm determine what is called the *algorithms complexity*. Specifically, the complexity of an algorithm is determined by the number of operations and memory usage required to complete the program scales with the size of the input of the program. The size of the input is conveniently defined here to be the number of bits needed to represent that number in binary. Computer Scientists have grouped problems into *complexity classes*, below are some of the more well know.

- *P: Polynomial time, the running time of the given algorithm is in the worst case some polynomial in the size of the input.*

- *NP: Nondeterministic polynomial time, a candidate for an answer can be verified as a correct answer or not in polynomial time.*

- *NP-complete: A set of problems for which if any can be solved in polynomial time, P equals NP.*

Problems, which can be solved in polynomial time or less, are generally deemed to be *tractable*. Problems, which require more than polynomial time are usually considered to be *intractable*, for example an algorithm which would take 2^n operations for an input size of n would be considered intractable, as the number of operations grows exponentially with the input size, not polynomial. In general, if an algorithm requires polynomial time or less to compute it is considered to be *tractable*, and if not it is considered to be *intractable*. Great attention was paid by Turing and others to attempt to make their models of computing mathematical abstractions, harboring no hidden assumptions about the computing machinery itself. For many years it appeared that these models were indeed not based in any assumptions as to the nature of the computer, but this is not so.

The number of atoms needed to represent a bit in memory has been decreasing exponentially since 1950. Likewise, the numbers of transistors per chip, clock speed, and energy dissipated per logical operation have all followed their own improving exponential trends. Despite these fantastic advances, the manner in which all computers function is essentially identical. At the current rate in the year 2020, one bit of information will requite only one atom to represent it. The problem is that at that level of miniaturization the behavior of the components of a computer will become dominated by the principles of quantum physics.

With the size of components in classical computers, shrinking to where the behavior of the components may soon be dominated more by quantum physics than classical physics researchers have begun investigating the potential of these quantum behaviors for computation. When components are able to function in quantum, unexpectedly it seems the possibility to obtain a more powerful computer than any classical computer can be.

A *quantum computer* (if one could be constructed) is more powerful than Turing machines, in the sense that there are problems, which are solvable on a quantum computer, which are not solvable on a Turing machine. The quantum computer is more powerful than the Turing machine and hence any classical computer because it can do things, which were thought to be physically impossible when Church's thesis was written. Church's thesis is still valid for any computing machine, which functions in a purely classical manner.

The laws of quantum mechanics could support new types of algorithms (*quantum algorithms*) Deutsch was able to define new complexity classes (Deutsch 1985) and establish a new hierarchy. However, it was not recognized until recently that the class of problems that can be solved in polynomial time with a quantum algorithm.

- *QP: A set of problems for which the best classical algorithm runs exponentially.*

In other words, quantum computers are able to perform (in polynomial time) certain computations for which no polynomial time algorithm is known.

2 Digital Systems

Before we can appreciate the meaning and implications of *digital systems*, it is necessary to look at the nature of *analog systems*. The world is *analog* is derived from the same root as the noun *analogy* and means a quantity that is related to, or corresponds to, another quantity. In a digital system, all variables and constants must take a value chosen from a set of values called an *alphabet*. In decimal arithmetic, we have an alphabet composed of the symbols 0 through 9. Other digital systems are Morse, Braille, semaphore and the day of the week.

An advantage of digital representation of information is that a symbol may be distorted, but as long as the level of distortion is not sufficient for one symbol to be taken for another, the original symbol may be recognized and reconstituted.

The alphabet selected for digital computers has two symbols, 0 and 1. The advantage of such an alphabet is that the symbols can be made unlike each other as possible to give the maximum discrimination between the two values. A single binary digit is known as a *bit* (*bi*nary dig*it*) and is the smallest unit of information possible.

2.1 Number Systems

The number system we use every day is a *positional number system*. In such a system, any number is represented by a string of digits in which the position of each digit has an associated weight. The value of a given number, then, is equivalent to the weighted sum of all its digits. A number written $a_{n-1}...a_1a_0.a_{-1}a_{-2}...a_{-m}$ when expressed in positional notation in the *base d* is defined as

$$a_{n-1}d^{n-1} + ... + a_1d^1 + a_0d^0 + a_{-1}d^{-1} + ... + a_{-m}d^{-m} = \sum_{i=-m}^{n-1} a_i d_i$$

where there are n digits to the left of the point, known as the *radix point*, and m digits to its right. Within a positional number system, the representation of every number is unique, except for possible leading and trailing zeros. Note that the leftmost digit in such a number system is called the most-significant digit (MSD), and the rightmost is the least-significant digit (LSD).

Since digital system use binary digits, we use a binary base to represent any number in a digital system. The general form of such a binary number is $b_{n-1}...b_1 b_0 .b_{-1} b_{-2}...b_{-m}$ and its value is equivalent to

$$B = \sum_{i=-m}^{n-1} b_i (2)^i$$

Similarly, to a decimal point in a decimal number, the radix point in a binary number is called the *binary point*.

2.2 Hexadecimal Numbers

The hexadecimal number system uses base 16. The hexadecimal system needs to express 16 different values, so it supplements the decimal digits 0 through 9 with the letters A through F. From our discussion of positional number systems, we can easily recognize the importance of base 10, which use in everyday life, and base 2, which is used by digital systems to process numbers.

The general form of such hexadecimal numbers is $h_{n-1}...h_1 h_0 .h_{-1} h_{-2}...h_{-m}$ and its value is equivalent to

$$H = \sum_{i=-m}^{n-1} b_i (16)^i$$

The binary integers 0 through 1111 and their binary-coded decimal and hexadecimal equivalents are given below

DECIMAL	HEXADECIMAL	BINARY
0	0	0000
1	1	0001
2	2	0010
3	3	0011
4	4	0100
5	5	0101
6	6	0110
7	7	0111
8	8	1000
9	9	1001
10	A	1010
11	B	1011
12	C	1100
13	D	1101
14	E	1110
15	F	1111

Hexadecimal numbers are often used to provide convenient shorthand representations for binary numbers, reducing the need for lengthy, indecipherable strings. The hexadecimal number system is quite popular, because it can easily be converted to and from binary, and because standard byte, word, double-word data can be documented efficiently by hexadecimal digits.

As a rule we cannot covert a number representation in one base into representation in another base simply by substituting numbers in one base from their equivalent in another base representation; this works only when both bases are powers of the same number. When this is not the case, we must use more complex conversion procedures that require arithmetic operations.

2.3   Coding Theory

In the world of digital systems, there are many different codes, each one the best suited to the particular job for which it was designed. A particularly widespread binary code is called BCD or *binary-coded decimal.* BCD numbers accept the inevitability of two-state representation by coding the individual decimal digits groups of four bits as given below

DECIMAL	BCD
0	0000
1	0001
2	0010
3	0011
4	0100
5	0101
6	0110
7	0111
8	1000
9	1001

The BCD is often called pure binary, natural binary, or 8421 weighted binary. The 8, 4, 2, and 1 represent the *weightings* of each of the columns in the positional code. Many other positional codes do not have a natural binary weighting. Some codes are called *unweighted* because the value of a bit does not depend on its *position* in a number. Each of these codes has special properties that make it suitable for a specific application.

Within a digital system, an error can be caused only by temporary or permanent physical failures and can be defined as the difference between transmitted and received data. To detect such errors, we need to secure the data with the help of error-detecting code. The goal of *coding theory* is to represent digital information in a form that allows for the recovery of the original data from its corrupted form, if the number of errors is not too large. This requires that some redundancy be incorporated into the stored information.

Information is stored and transmitted as a stream of *letters* from a chosen *alphabet A*. Most popular is the *binary alphabet* $A = \{0,1\}$. More generally, $A = \{0,1,2,...,p-1\}$ with addition and multiplication mod p (where p is a prime number) is popular because A is a field. In this case

14

$$A^n = \{(a_1, a_2, ..., a_n) : a_i \in A\}$$

is an n-dimensional vector space over A. A *word* of length n is a string of n characters from the alphabet A. If $A = |q|$ then there are q^n words of length n. These are identified with the vectors of A^n. A *code of length* n is a subset $C \subseteq A^n$. Elements of C are *code words* and C is a *binary code* if $A = \{0,1\}$.

❖ Parity Check Code Example:

The following binary code $C_p = \{0000, 00011, ... 11110\}$ of length 5 is formed by appending a *parity check bit* to the end of each message word.

MESSAGE WORD	CODE WORD
0000	00000
0001	00011
0010	00101
0011	00110
0100	01001
0101	01010
0110	01100
0111	01111
1000	10001
1001	10010
1010	10100
1011	10111
1100	11000
1101	11011
1110	11101
1111	11110

Using the code C_p, we can detect up to one-bit error during transmission, but we cannot correct any errors.

❖ 3-Repetitions Code Example:

The following binary code

$$C_3 = \{000000000000, 000100010001, \ldots 111111111111\}$$

of length 12 is formed by repeating each message word three times.

MESSAGE WORD	CODE WORD
0000	000000000000
0001	000100010001
0010	001000100010
0011	001100110011
0100	010001000100
0101	010101010101
0110	011001100110
0111	011101110111
1000	100010001000
1001	100110011001
1010	101010101010
1011	101110111011
1100	110011001100
1101	110111011101
1110	111011101110
1111	111111111111

Using this code, we can correct up to one-bit error during transmission.

This gain comes at a price: C_3 has information rate $4/12 = 1/3$, lower than the information rate of C_p which is $4/5$. The information rate of a binary code is the ratio of the number of significant bits of information in each word, to the total length of each word. More generally for an alphabet of size $A = |q|$, the *information rate* of a code C of length n over A is $\log_q |C|/n$. We seek codes with high information rate, *and* high error-correcting capability.

Richard W. Hamming (1915–1998), was a pioneer in computer design and error-correcting codes. The *Hamming distance* between two words $x, y \in A^n$, denoted $d(x, y)$, is the number of position in which they differ.

❖ Example:

$$d(10010, 00111) = 3$$

The *minimum distance* of a code $C \subseteq A^n$ is the minimum of $d(x, y)$ for all $x \neq y$ in the code C. In 1950, Hamming proposed a general method for constructing error-correcting codes. Hamming codes are possibly the simplest class of error-detecting and correcting codes that can be applied to a single code word.

For every integer m there is a (2^m - 1) bit Hamming code which contains m parity bits and (2^m - 1 - m) information bits. Within this Hamming code, the parity bits are intermixed with the information bits as follows: If we number the bit positions from 1 to 2^m - 1, the bits in position 2^k, where $0 \leq k \leq m - 1$, parity bits and the bits in the remaining positions are information bits. Thus, parity bits are always in position 2^0 through 2^{m-1}.

In general, the value of each parity bit is chosen so that the total number of 1's in a specific group of bits positions is even, and these groups are chosen so that no information bit is covered by the same combination of parity bits. It is this arrangement that gives the code its correcting capability. More precisely, for each parity bit in position 2^k, its corresponding group of information bits includes all those bits in the position whose binary representation has a 1 in position 2^k.

Hamming code described above can detect and correct a single error. By adding a future check bit, we can create a Hamming code that can detect two errors and correct one.

George Boole was an English mathematician (1815 - 1864) who developed a mathematical analysis of logic and published it in his book *"An investigation of the laws of thought"* in 1854. In 1938, Claude Shannon published a paper entitled *"A symbolic analysis of relay and switching circuits"* which applied Boolean algebra to switching circuits using relay. In digital systems, Boolean algebra is used to design digital circuits and to analyze their behavior.

Boolean algebra consists of a set of *elements S*, a set of *functions f* that operate on members of *S*, and a set of basic laws called *axioms* that define the properties of *S* and *f*. The set of elements making up a Boolean algebra have two possible values, 0 or 1. These elements may be variables or they may be literals (i.e. constants) that have fixed values of 0 or 1. A Boolean algebra with n variables has a set of 2^n possible sets of values of these variables.

Only three functions or operations are permitted in Boolean algebra. The first two are the logical *OR* represented by a plus '+', and the logical *AND* represented by a dot '·'. The third operation permitted in Boolean algebra is that of negation (*NOT*) or complementation. The complement of 0 (i.e. $0'$) is 1, and the complement of the 1 (i.e. $1'$) is 0. The priority of an *AND* operator is higher than that of an *OR* operator. The effect of the three operations *OR*, *AND*, *NOT*, is illustrated by means of the truth table given below.

NOT	AND	OR
$0' = 1$	$0 \cdot 0 = 0$	$0 + 0 = 0$
$1' = 0$	$0 \cdot 1 = 0$	$0 + 1 = 1$
	$1 \cdot 0 = 0$	$1 + 0 = 1$
	$1 \cdot 1 = 1$	$1 + 1 = 1$

These rules may be extended to any number of variables. Boolean variables obey similar commutative, distributive, and associative laws as the variables of conventional algebra. If and only if for every $(x_0, x_1) \in S$, $x_0 \cdot x_1$, $x_0 + x_1$, x_0', x_1', also belong to the set of Boolean elements. This axiom is called the enclosure property and implies that Boolean operations on Boolean variables or constants always yield Boolean results.

The basic axioms of Boolean algebra governing variables, operators and constants are given below.

AND	OR	NOT
$0 \cdot x = 0$	$0 + x = x$	$(x')' = x$
$1 \cdot x = x$	$1 + x = 1$	
$x \cdot x = x$	$x + x = x$	
$x \cdot x' = 0$	$x + x' = 1$	

These equations may be proved by substituting all the possible value for x (i.e. 0 or 1). One of each pair of axioms can be obtained from the other by using an important property of Boolean algebra called the *duality principle*. This principle states that any algebraic equality derived from these axioms will still be valid whenever the *OR* and *AND* operators, and identity elements 0 and 1, have been interchanged.

In addition to the axioms that are given in the definition of a Boolean algebra, we can also derive additional laws, called *theorems* of Boolean algebra. The lists of six basic theorems of Boolean algebra are given below.

Theorem 1 (idempotency)	$x_0 + x_0 = x_0$	$x_0 \cdot x_0 = x_0$
Theorem 2	$x_0 + 1 = 1$	$x_0 \cdot 0 = 0$
Theorem 3 (absorption)	$x_1 \cdot x_0 + x_0 = x_0$	$(x_1 + x_0) \cdot x_0 = x_0$
Theorem 4 (involution)	$(x_0)' = x_0$	
Theorem 5 (associativity)	$(x_0 + x_1) + x_2 = x_0 + (x_1 + x_2)$	$x_0 \cdot (x_1 \cdot x_2) = (x_0 \cdot x_1) \cdot x_2$
Theorem 6 (De Morgan's law)	$(x_0 + x_1)' = x_0' \cdot x_1'$	$(x_0 \cdot x_1)' = x_0' + x_1'$

These theorems are particular useful when we perform algebraic manipulation of Boolean expressions. Axioms are given and need no proof. Theorems must be proven, either from axioms or from other theorems that have already been proven. Since the two-valued Boolean algebra has only two elements, we can also show the validity of these theorems by using truth tables. A truth table is constructed for each side of the equation that is stated in the theorem. Then both sides of the equation are checked to see that they yield identical results for all possible combinations of variable values.

2.5   Boolean Functions

In general, functions can be defined as algebraic expressions that are formed from variables, operators, parentheses, and an equal sign. More specifically, Boolean expressions are formed from binary variables and the Boolean operators *AND*, *OR*, and *NOT*.

When we compute the values of Boolean expressions, we must adhere to a specific order of computation: namely, *NOT*, *AND*, and *OR*. Binary variables in these expressions can be take only two values, 0 and 1. For a given value of the variables, then, the value of the function is either, 0 and 1. The Boolean expression can be characterized as having *OR terms* and *AND terms*. Each term contains literals, where a *literal* indicates a variable or its complement. The number of terms and literals is usually used as a measure of the expression's complexity and frequently as a measure of its implementation cost. Consider, for example, the Boolean function $f(x_0, x_1, x_2) = x_0 \cdot x_1 + x_0 \cdot x_1' \cdot x_2' + x_0' \cdot x_1 \cdot x_2$ can be characterized as having one *OR* term and three *AND* terms.

Any Boolean function can also be defined by a truth table, which lists the value the function has for each combination of its variables values. As a rule, the truth table for any Boolean function of n variables has 2^n rows, which represent every possible combination of variable values, so that value entered in the function column in each row is the value of the function for that particular combination of variable values.

2.6 Canonical Forms

In this section, we show how truth tables can be converted into algebraic expressions. For an n-variable function, each row in the truth table represents a particular assignment of binary values to those n variables. We can define a Boolean function, usually called *minterm*, which is equal to 1 in only one row in the truth table and 0 in all other rows. Since there are 2^n different rows in the truth table, there are also 2^n minterms for any n variables. Each of these minterms, denoted as m_i, can also be expressed as a product, or *AND* term, of n literals, in which each variable is complemented if the value assigned to it is 0, and uncomplemented if it is 1. More formally, this means that each minterm m_i can be defined as follows.

Let $i = b_{n-1}...b_0$ be a binary number between 0 and $2^n - 1$, which represents an assignment of binary values to the n binary variables x_j such that $x_j = b_j$ for all j where $0 \leq j \leq n - 1$. In this case a minterm of n variables $x_{n-1}, x_{n-2}, ..., x_0$, could be represented as

$$m_i(x_{n-1}, x_{n-2}, ..., x_0) = a_{n-1}a_{n-2}...a_0$$

where for all k such that $0 \leq k \leq n - 1$,

$$a_k = \begin{cases} x_k & \text{if } b_k = 1 \\ x_k' & \text{if } b_k = 0 \end{cases}$$

The unique algebraic expression for any Boolean function can be obtained from its truth table by using an *OR* operator to combine all minterms for which the function is equal to 1. In other words, any Boolean function can be expressed as a sum of its minterms

21

$$f(x_{n-1},...,x_0) = \sum_{F?=1} m_i$$

for which the function is equal to 1. The *OR*ing of minterms is called a sum because of its resemblance to summation.

Similarly to an *AND* term of n literals being called a minterm, an *OR* term of n literals is called a *maxterm*. A maxterm can be defined as a Boolean function that is equal to 0 in only one row of its truth table and 1 in all other row. Each maxterm, M_i, can be expressed as a sum, of *OR* term, of n literals in which each variable would be uncomplemented if the value assigned to it is 0, and complemented if it is 1. Note that each maxterm is the complement of its corresponding minterm, and vice versa, $(m_i)' = M_i$ and $(M_i)' = m_i$.

The unique algebraic expression for any Boolean function can be obtained from its truth table by using an AND operator to combine all the maxterms for which the function is equal to 0. Any Boolean expression can be expressed as a product of its maxterms

$$f(x_{n-1},...,x_0) = \prod_{F?=0} M_i$$

for which the function is equal to 0. The *AND*ing of maxterms is called a product because of its resemblance to multiplication.

Any Boolean function that is expressed as a sum of minterms or as a product of maxterms is said to be in its canonical form. These two canonical forms provide unique expression for any Boolean function defined by truth table.

2.7   Logic Gates

To implement the Boolean functions we construct *logic circuit*, which contains one or more *logic gates*. Each logic gates performs one or more Boolean operation (*AND, OR, NOT*). The collection of logic gates that we use in constructing logic circuits is called the *gate library*, and the gates in the library are called *standard gates*.

In particular case of two variables we have $2^2 = 4$ possible different combinations. We can associate a different function with each of these $4^2 = 16$ values to create all possible functions of two variables. In general, n variables have $(2^2)^n$ functions which are given below

NAME	FUNCTION VALUE FOR $(x_0, x_1) =$ 00 01 10 11				ALGEBRAIC EXPRESSION $y_i = f(x_0, x_1)$
Zero	0	0	0	0	$y_0 = 0$
AND	0	0	0	1	$y_1 = x_0 \cdot x_1$
Inhibition	0	0	1	0	$y_2 = x_0 \cdot x_1'$
Transfer	0	0	1	1	$y_3 = x_0$
Inhibition	0	1	0	0	$y_4 = x_0' \cdot x_1$
Transfer	0	1	0	1	$y_5 = x_1$
XOR	0	1	1	0	$y_6 = x_0 \cdot x_1' + x_0' \cdot x_1$
OR	0	1	1	1	$y_7 = x_0 + x_1$
NOR	1	0	0	0	$y_8 = (x_0 + x_1)'$
XNOR	1	0	0	1	$y_9 = x_0 \cdot x_1 + x_0' \cdot x_1'$
Complement	1	0	1	0	$y_{10} = x_1'$
Implication	1	0	1	1	$y_{11} = x_0 + x_1'$
Complement	1	1	0	0	$y_{12} = x_0'$
Implication	1	1	0	1	$y_{13} = x_0' + x_1$
NAND	1	1	1	0	$y_{14} = (x_0 \cdot x_1)'$
One	1	1	1	1	$y_{15} = 1$

Two of these functions called *NOR* and *NAND* are intrinsically better than others are because this function (*De Morgan's law*) can provide *AND*, *OR* and *NOT* Boolean operation.
Based of technological criteria only eight functions was selected to be implemented as *standard gate*, namely, the *Complement*, *Transfer*, *AND*, *OR*, *NAND*, *NOR*, *XOR*, and *XNOR* functions. In

table given below, we show the graphic symbols and Boolean expression of each of these eight gates.

NAME	GRAPHIC SYMBOL	FUNCTIONAL EXPRESSION
Inverter		$y = x_0'$
Driver		$y = x_0$
AND		$y = x_0 \cdot x_1$
OR		$y = x_0 + x_1$
NAND		$y = (x_0 \cdot x_1)'$
NOR		$y = (x_0 + x_1)'$
XOR		$y = x_0 \cdot x_1' + x_0' \cdot x_1$
XNOR		$y = x_0 \cdot x_1 + x_0' \cdot x_1'$

The function of the *inverter* is to complement the logic value of its input; we place a small circle at the output of its graphic symbol to indicate this *logic complementation*. We also use a triangle symbol to designate a *driver* circuit, which implements the *transfer function* by replicating the input value at its output. A diver is equivalent to two inverters that are connected in cascade, so that the output of the first inverter serves as the input of the second.

24

The *AND* and *OR* gates are used to implement the Boolean operators *AND* and *OR*, whereas the *NAND* and *NOR* gates are used to implement those functions that are the complement of *AND* and *OR*. A small circle is used at each of these outputs, which indicate this complementation. The *NAND* and *OR* gates are used extensively and are far more popular than the *AND* and *OR* gates, simply because their implementation is more simple in comparison with *AND* and *OR* gates. Note that the *XNOR* gate is the complement of *XOR* gate, as indicated by the small circle on its output line. When we are implementing Boolean functions with this basic gate library, we usually try to find the Boolean expression, of the canonical form of the function, which will best satisfy a given set of design requirements.

2.8   Finite State Machine

Each logic circuit we have encountered up to this point has been a *combinational circuit* whose output is a function of its input only. That is, given knowledge of a combinational circuit's inputs together with its Boolean functions, we can always calculate the state of its outputs. Circuits, whose outputs depend not only on their current inputs, but also on their past inputs, are called *sequential circuits*. Even if we know the structure of a sequential circuit (i.e. its Boolean function) and its current input, we cannot determine its output state without knowledge of its past history (i.e. its past internal state).

The basic building block of sequential circuits is *bistable*, just as the basic building block of the combinational circuit is the gate. A bistable is so called because, for a given input, its output can remain in one of two stable states indefinitely. For a particular set of inputs, the output may assume either a logical zero or a logical one, the actual value depending on the previous inputs. Such a circuit has the ability to remember what has happened to it in the past and is therefore a form of *memory element*.

25

The finite-state machine (FSM) can be defined abstractly as the quintuple $\langle S, I, O, f, h \rangle$, where S, I and O represent a set of states, set of inputs, and a set of outputs, respectively, and f and h represent the next-state and output functions. The next-state function f is defined abstractly as a mapping $S \times I \rightarrow S$. In other words, f assign to every pair of state and input symbols another state symbol. The FSM model assumes that time is divided into uniform intervals and that transitions from one state to another occur only at the beginning of each time interval. Therefore, the next-state function f defines what the state of the FSM will be in the next time interval given the state and input values in the present interval.

The output function h determines the output value in the present state. There are two different types of finite-state machine, which correspond to two different definitions of the output function h. One type is Moore FSM, for which h is defined as a mapping $S \rightarrow O$. In other words, an output symbol is assigned to each state of the FSM. The other type is Mealy FSM, for which h is defined as the mapping $S \times I \rightarrow O$. In this case, a pair of state and input symbols defines an output symbol in each state.

The FSM can model any sequential circuit with k input signals $A_1, A_2, ..., A_k$, m bistables (memory register) $Q_1, Q_2, ..., Q_m$, and n output signals $Y_1, Y_2, ..., Y_n$, as show below

$$I = A_1 \times A_2 \times ... \times A_k$$
$$S = Q_1 \times Q_2 \times ... \times Q_m$$
$$O = Y_1 \times Y_2 \times ... \times Y_n$$

For such a sequential circuit, S, I, and O are cross products of bistables or signals. Thus each element in S, I, and O is represented by a string of 1's and 0's. An external clock (Clk) signal, which driven the bistables, defines the time intervals, called *clock cycles*.

The first simple abstract computational device (Turing 1937) described by Alan Turing intend to help investigate the extent and limitations of what can be computed. Turing, writing before the invention of the modern digital computer, was interested in the question of what it means to be computable. Intuitively a task is computable if one can specify a sequence of instructions which when followed will result in the completion of the task. Such a set of instructions is called *algorithm* of the task. This intuition must be made precise by defining the capabilities of the device that is to carry out the instructions. Turing proposed a class of devices that came a formal notion of computation that we will call *Turing-computability*.

Turing machine consists of an infinitely long *tape* with symbols (chosen from some finite set) written at regular intervals. A pointer marks the current position and the machine is in one of a finite set of *internal states*. The action of a Turing machine is determined completely by:

o   the current state of the machine,
o   the symbol in the cell currently being scanned by the head and
o   a table of transition rules, which serve as the *program* for the machine.

Each transition rule is a 4-tuple:

$$\langle State_0, Symbol, State_{next}, Action \rangle$$

which can be read as saying "if the machine is in state $State_0$ and the current cell contains $Symbol$ then move into state $State_{next}$ taking $Action$". In other word, at each step the machine reads the symbol at the current position on the tape. For each combination of current state and symbol read, a program specifies the new state and either a symbol to write to the tape or a direction to move the pointer (left or right) or to halt.

Without loss of generality, the symbol set can be limited to just 0 and 1 and the machine can be restricted to start on the leftmost 1 of the leftmost string of 1s with strings of 1s being separated by a single 0. The tape may be infinite in one direction only, with the understanding that the machine will halt if it tries to move off the other end.

All computer instruction sets, high level languages and computer architectures, including parallel processors, can be shown to be equivalent to a Turing Machine and thus equivalent to each other in the sense that any problem that one can solve, any other can solve given sufficient time and memory. Turing generalized the idea of the Turing Machine to a *Universal Turing Machine*, which was programmed to read instructions, as well as data, off the tape, thus giving rise to the idea of a general-purpose programmable computing device. This idea still exists in modern computer design with low-level microcode, which directs the reading and decoding of higher-level machine code instructions.

3 Quantum Systems

Quantum physics arose from the failure of classical physics to offer correct predictions on the behavior of photons and other elementary particles. The study of information encoded in *quantum systems* as a coherent discipline began to emerge in the 1980's. Many of the central results of classical information theory have quantum analogs that have been discovered and developed recently. Information is something that is encoded in the state of a physical system; a computation is something that can be carried out on an actual physically realizable device. Therefore, the study of information and computation should be linked to the study of the underlying physical processes.

An important distinction needs to be made between quantum mechanics, quantum physics and *quantum computing*. Quantum mechanics is a mathematical model of the physical world. Just as classical physics uses calculus to explain nature, quantum physics uses quantum mechanics to explain nature. Just as classical computers can be thought of in Boolean algebra terms, quantum computers are reasoned about with quantum mechanics.

The quantum information are important, but the really deep way in which quantum information differs from classical information emerged from the work of John Bell (1964), who showed that the predictions of quantum mechanics cannot be reproduced by any local hidden variable theory. Bell showed that quantum information could be encoded in non-local correlations between the different parts of a physical system, correlations with no classical counterpart.

3.1 Hilbert Space

In general, the mathematical term *space* refers to a something, which depends on many independent coordinates, which can be defined by a set of perpendicular axes, one for each independent variable. For quantum computation, a complex vector space C^n is used not R^n. The numbers which take the form $c = a + b \cdot i$, where a and b are real numbers and $i = \sqrt{-1}$ are complex numbers. The complex conjugate is defined as $c^* = a - b \cdot i$, where $c, c^* \in C^n$. Multiplication of two complex numbers $c_1 = a_1 + b_1 i$ and $c_2 = a_2 + b_2 \cdot i$ is defined to be:

$$c_1 \cdot c_2 = a_1 \cdot a_2 - b_1 \cdot b_2 + i(a_1 \cdot b_2 + a_2 b_1)$$

❖ Example

$$c = 2 + 3 \cdot i \quad \text{and complex conjugate} \quad c^* = 2 - 3 \cdot i$$

$$c \cdot c^* = (2 \cdot 2) - (3 \cdot -3) + i((2 \cdot -3) + (2 \cdot 3)) = 13$$

The vectors of complex space will be denoted $|\psi\rangle$, called a *ket vector*. The ket vector

$$|\psi\rangle = \begin{pmatrix} c_0 \\ c_1 \\ . \\ . \\ . \end{pmatrix}$$

is a list of numbers which contain information about the projection of the state vector onto its base states. The term ket and this notation come from the physicist Paul Dirac who wanted a concise shorthand way of writing formulas that occur in quantum physics. These formulas frequently took the form of the product of a *row vector* with a *column vector*. Thus, he referred to row vectors as *bra vectors* represented as

$$\langle \psi | = (c_0^*, c_1^*, ...)$$

The inner product of a *bra vector* $\langle \varphi |$ and a *ket vector* $| \psi \rangle$ would be written $\langle \varphi | \psi \rangle$, and would be referred to as a *bracket*.

❖ Example:

$$| \varphi \rangle = \begin{bmatrix} 3 \\ 6i \end{bmatrix}, \quad \langle \varphi | = [3 \quad -6i], \quad | \psi \rangle = \begin{bmatrix} 4 \\ 5 \end{bmatrix}$$

$$\langle \varphi | \psi \rangle = [3 \quad -6i] \begin{bmatrix} 4 \\ 5 \end{bmatrix} = 12 - 30i$$

A Hilbert space is an inner product space, which is complete with respect to the norm. The inner product $\langle \varphi | \psi \rangle$ meets the following conditions:

- Positivity: $\langle \varphi | \psi \rangle \geq 0$ for $| \psi \rangle \neq 0 \Rightarrow \langle \psi | \psi \rangle > 0$
- Linearity: $\langle \varphi | (a | \psi_1 \rangle + b | \psi_2 \rangle) = a \langle \varphi | \psi_1 \rangle + b \langle \varphi | \psi_2 \rangle$
- Skew symmetry: $\langle \varphi | \psi \rangle = \langle \psi | \varphi \rangle^*$

The Hilbert space is complete in norm $\| \psi \| = \sqrt{\langle \psi | \psi \rangle}$. By complete we mean that every sequence of vectors $| \psi_n \rangle$, with the property that:

$$\forall \varepsilon > 0 \ \exists N > 0 \ \forall n > N, \left\| | \psi_n \rangle - | \psi \rangle \right\| < \varepsilon$$

has a limit in V .

In other words, a Hilbert Space is a special kind of space, it has the properties that it is a complex vector space, and it is a linear vector space. The fact that it is a complex vector space means that the lengths of the vectors within the space are described with complex numbers. The fact that it is a linear vector space means that you may add and multiply vectors that lie in a given Hilbert Space and

31

the resulting vector will still lie within that Hilbert Space. In addition, a very useful fact about Hilbert spaces is that all Hilbert spaces of a given dimension are *isomorphic*.

3.2   Qubit

The most fundamental building block of a classical computer is the bit. A bit is capable of storing one piece of information; it can have a value of either 0 or 1. Any amount of information can be encoded into a list of bits. About 10^{10} atoms are currently used to represent one bit of information in a classical computer.

In a quantum system, information is represented as the common quantum state of many subsystems. Any two state quantum subsystems such as the ground and excited state of an ion, the spin of an electron, or the polarization of a photon can be used to store one bit of information. We could call it a *qubit*, to denote that it is analogous in some ways to a bit in a classical computer. However, if we were to try to use this qubit as a classical bit, one that is always in the 0 or 1 state, we would fail. We would be trying to apply classical physics on a scale where it simply is not applicable.

We wish to know exactly how the behavior of the qubit differs from a classical bit. Recall that a classical bit can store either a 1 or a 0, and when measure the value observed we will get always the value stored. Quantum physics states that when we measure the state of qubit we will determine that it is in 1 or 0 state. In this manner, our qubit is not different from a classical bit. The differences between the qubit and the bit come from what sort of information a qubit can store when it is not being measured.

According to quantum physics, we may describe that state of this qubit by a state vector in a Hilbert Space. Therefore, Hilbert Space for a single qubit will have two perpendicular axes, one corresponding to the qubit being in the 1 state, and the other to the qubit being in the 0 state. These states, which the vector can be measured to be, are referred to as *eigenstates*. The vector which exists somewhere in this space which represents the state of our qubit is called the *state vector*. The projection of the state vector onto one of the axes shows the contribution of that axe's eigenstate to the whole state. This means that in general, the state of the qubit can be any combination of two *pure* states interpreted as 0 and 1. In this manner a qubit it totally unlike a bit, for a bit can exist in only the 0 or 1 state, but the qubit can exist in any combination of the 0 and 1 state, and is only constrained to be in the 0 or 1 state when we measure the state.

According to quantum physics, a quantum system can exist in a mix of all of its allowed states simultaneously. This is the *principle of superposition*, and it is the source to the *power* of the quantum computer. While the physics of superposition is not simple at all, mathematically it is not difficult to characterize this kind of behavior.

This qubit, which behaves in a quantum manner, could be the fundamental building block of a *quantum computer*. It is fundamentally different from classical bit because it can exist in any superposition of the 0 and 1 state when it is not being measured.

Associated to any isolated physical system is a Hilbert space known as the state space of the system. The system is completely described by its state vector, which is a unit vector in the system's state space.

We could refer to its state in the following manner. Let $|\varphi_0\rangle$ and $|\varphi_1\rangle$ be orthonormal basis for the space. Let $|\psi\rangle$ be the total state of our state vector, and let w_1 and w_0 be the complex numbers that weight the contribution of the base states to our total state, then in general:

$$|\psi\rangle = w_0|\varphi_0\rangle + w_1|\varphi_1\rangle = \begin{pmatrix} w_0 \\ w_1 \end{pmatrix}$$

At this point, it should be remembered that w_0 and w_1, the weighting factors of the base states are complex numbers and that when the state of $|\psi\rangle$ is measured, we are guaranteed to find it to be in either the state:

$$0|\varphi_0\rangle + w_1|\varphi_1\rangle = \begin{pmatrix} 0 \\ w_1 \end{pmatrix}$$

or the state

$$w_0|\varphi_0\rangle + 0|\varphi_1\rangle = \begin{pmatrix} w_0 \\ 0 \end{pmatrix}$$

For example, $|\varphi_0\rangle = |0\rangle$ and $|\varphi_1\rangle = |1\rangle$. Making this more concrete one might imagine that $|1\rangle$ is being represented by an up-spin while $|0\rangle$ by a down-spin. The key is there is an abstraction between the technology (spin state or other quantum phenomena) and the logical meaning.

Note that, our state vector is a unit vector in a Hilbert space,

$$\langle\psi|\psi\rangle = 1 \text{ or } w_0 w_0^* + w_1 w_1^* = 1$$

the lengths of the vectors are complex numbers. This normalization condition is not a property of quantum mechanics but rather of probability theory.

3.3   Quantum Register

Just as a memory register in a classical computer is an array of bits, a *quantum memory register* is composed of several qubits. In an n state system, our Hilbert Space has n perpendicular axes, or eigenstates, which represent the possible states that the system can be measured in. As with the two state systems, when we measure a

n state quantum system, we will always find it to be in exactly one of the n states, and not a superposition of the n states. The system is still allowed to exist in any superposition of the n states while it is not being measured.

Mathematically if two, state quantum system with coordinate axes $|\varphi_0\rangle$ and $|\varphi_1\rangle$ can be fully described by:

$$|\psi\rangle = w_0|\varphi_0\rangle + w_1|\varphi_1\rangle = \begin{pmatrix} w_0 \\ w_1 \end{pmatrix}$$

Then an n state quantum system with coordinate axes $|\varphi_0, \varphi_1, ..., \varphi_{n-1}\rangle$ can be fully described by:

$$|\psi\rangle = \sum_{k=0}^{n-1} w_k|\varphi_k\rangle$$

where it is understood that w_k refers to the complex weighting factor for the $k^{'th}$ eigenstate.

In general, a quantum system with n base states can be represented by the n complex numbers w_0 to w_{n-1}. When this is done, the state may be written as:

$$|\psi\rangle = \begin{pmatrix} w_0 \\ w_1 \\ . \\ . \\ . \\ w_{n-1} \end{pmatrix}$$

Using this information, we can construct a quantum memory register out of the qubits described in the previous section. We may store any number n in the quantum memory register as long as we have enough qubits, just as we may store any number n in a classical register as long as we have enough classical bits to represent that number. The state of the quantum register with n states is give by the formula above. Note that, in general a quantum register composed of n qubits requires 2^n complex numbers to completely describe its state. A n qubit register can be measured to be in one of 2^n states, and each state requires one complex number to represent the projection of that total state onto that state. In contrast, a classical register composed of n bits requires only n integers to fully describe its state.

❖ Example:

$$|\psi\rangle = \begin{pmatrix} 0.3536 \\ -0.2500 - 0.2500i \\ 0 + 0.3536i \\ 0.2500 - 0.2500i \\ -0.3536 \\ 0.2500 + 0.2500i \\ 0 - 0.3536i \\ -0.2500 + 0.2500i \end{pmatrix}$$

This means that one can store an exponential amount of information in a quantum register. Here we see some of the first hints that a quantum computer can be exponentially more powerful than a classical computer in some respects.

3.4   Evolution of quantum systems

Time evolution of a quantum state is unitary; it is generated by a self-adjoint operator, called the Hamiltonian of the system. In the Schrödinger picture of dynamics, the vector describing the system moves in time as governed by the Schrödinger equation

$$\frac{d}{dt}|\psi(t)\rangle = -iH|\psi(t)\rangle$$

where H is the Hamiltonian. We may re-express this equation, to first order in the infinitesimal quantity dt, as

$$|\psi(t+dt)\rangle = (1-iHdt)|\psi(t)\rangle$$

The operator $U(t) \equiv 1 - iHdt$ is unitary; because H is self-adjoint it satisfies $U^\dagger U = I$ to linear order in dt, where $U^\dagger = (U^T)^*$ is the Hermitian conjugate matrix (adjoint). Since a product of unitary operators is finite, time evolution over a finite interval is also unitary

$$|\psi(t)\rangle = U(t)|\psi(0)\rangle$$

In the case where H is t-independent; we may write $U = e^{-itH}$, where U a linear operator is.

❖ Example:

$$\text{If } A = \begin{bmatrix} 1 & 0+6i \\ 0+3i & 2+4i \end{bmatrix} \text{ then } A^T = \begin{bmatrix} 1 & 0+3i \\ 0+6i & 2+4i \end{bmatrix}$$

$$\text{and } A^\dagger = (A^T)^* = \begin{bmatrix} 1 & 0-3i \\ 0-6i & 2-4i \end{bmatrix}$$

In a classical computer, the logic gates do the processing of information. A logic gate maps the state of its input bits into another state according to a truth table. The corresponding quantum gate is implemented via a linear transformation, linear operator, or unitary operator U that evolves the basic state into the corresponding state according with the same truth table. Not every linear operator is a quantum gate, though. The action of a quantum gate must preserve the inner product of the underlying Hilbert space. An inner product preserving operator is called a unitary operator. Such operators are isomorphism of Hilbert spaces since the inner product provides all the structure.

The state of a quantum computer with n qubits is a point in a $2n$-dimensional Hilbert space C^n. The theoretical storage capacity therefore increases exponentially with the number of qubits. In other words, any computational step can be described as an operator $U : |\psi\rangle \rightarrow |\psi'\rangle$ over C^n or a subspace of C^n which transforms the input state $|\psi\rangle$ to the output state $|\psi'\rangle = U|\psi\rangle$.

❖ Example:

Let $|\psi\rangle = 1|0\rangle + 0|1\rangle = \begin{bmatrix} 1 \\ 0 \end{bmatrix}$

$U = \dfrac{1}{\sqrt{2}} \begin{bmatrix} 1 & 1 \\ 1 & -1 \end{bmatrix}$

$|\psi'\rangle = U|\psi\rangle = \dfrac{1}{\sqrt{2}} \begin{bmatrix} 1 & 1 \\ 1 & -1 \end{bmatrix} \begin{bmatrix} 1 \\ 0 \end{bmatrix} = \dfrac{1}{\sqrt{2}} \begin{bmatrix} 1 \\ 1 \end{bmatrix} = \dfrac{1}{\sqrt{2}}|0\rangle + \dfrac{1}{\sqrt{2}}|1\rangle$

The evolution of a closed quantum system is described by a unitary transformation. The state $|\psi\rangle$ of the system at time t_1 is related to the state $|\psi'\rangle$ of the system at time t_2 by a unitary operator U, which depends only on times t_1, and t_2.

❖ Example: U must be unitary, that is $U^\dagger U = I$

Let $U = \dfrac{1}{\sqrt{2}} \begin{bmatrix} 1 & 1 \\ 1 & -1 \end{bmatrix}$

$$U = U^{\dagger} \cdot U = \frac{1}{\sqrt{2}} \frac{1}{\sqrt{2}} \begin{bmatrix} 1 & 1 \\ 1 & -1 \end{bmatrix} \begin{bmatrix} 1 & 1 \\ 1 & -1 \end{bmatrix} = \frac{1}{2} \begin{bmatrix} 2 & 0 \\ 0 & 2 \end{bmatrix} = \begin{bmatrix} 1 & 0 \\ 0 & 1 \end{bmatrix}$$

The quantum computers can only perform reversible operations. Conceptually think of U as something you can apply to a quantum bit but you cannot conditionally apply it. The transform occurs without any regard to the current state of $|\psi\rangle$.

By a quantum computing device we mean a unitary transformation U that is the composition of finitely many local unitary transformations,

$$U = U_{n-1} U_{n-2} ... U_1 U_0$$

where $U_{n-1}, U_{n-2}, ..., U_1, U_0$ are local unitary transformations. Each U_j is called a computational step of the device.

3.5   Measurement

In a classical computer, the formal description of the inner state and the measurement of this state (i.e. the output of the program) is the same and given by the binary values of the concerned bits. Moreover, the inner state is not affected by the process of measurement (non destructive measurement). Not this is in the case of measurement process of quantum register.

Quantum measurements are described by a collection $\{M_m\}$ of measurement operators. These are operators acting on the state space of the system being measured. The index m refers to the measurement outcomes that may occur in the experiment. If the state of the quantum system is $|\psi\rangle$ immediately before the measurement then the probability that result m occurs is given by:

$$p(m) = \langle \psi | M_m^\dagger M_m | \psi \rangle$$

and the state of the system after measurement is:

$$\frac{M_m | \psi \rangle}{\sqrt{\langle \psi | M_m^\dagger M_m | \psi \rangle}}$$

The measurement operators satisfy the *completeness equation*:

$$\sum_m \langle \psi | M_m^\dagger M_m | \psi \rangle = I$$

The completeness equation expresses the fact that probabilities sum to one:

$$1 = \sum_m p(m) = \sum_m \langle \psi | M_m^\dagger M_m | \psi \rangle$$

❖ Example:
   Some important measurement operators are $M_0 = |0\rangle\langle 0|$
   and $M_1 = |1\rangle\langle 1|$

$$M_0 = \begin{bmatrix} 1 \\ 0 \end{bmatrix} [1 \ 0] = \begin{bmatrix} 1 & 0 \\ 0 & 0 \end{bmatrix}, \quad M_1 = \begin{bmatrix} 0 \\ 1 \end{bmatrix} [0 \ 1] = \begin{bmatrix} 0 & 0 \\ 0 & 1 \end{bmatrix}$$

Observe that $M_0^\dagger M_0 + M_1^\dagger M_1 = I$ and are thus complete.

❖ Example:

$$|\psi\rangle = a|0\rangle + b|1\rangle$$
$$p(0) = \langle \psi | M_0^\dagger M_0 | \psi \rangle$$

Note that $M_0^\dagger M_0 = M_0$, hence

$$p(0) = \langle \psi | M_0 | \psi \rangle = [a^* \ b^*] \begin{bmatrix} 1 & 0 \\ 0 & 0 \end{bmatrix} \begin{bmatrix} a \\ b \end{bmatrix} = [a^* \ b^*] \begin{bmatrix} a \\ 0 \end{bmatrix} = |a|^2$$

Hence, the probability of measuring $|0\rangle$ is related to its probability amplitude a by way of $|a|^2$. The state after measurement is related to the outcome of the measurement. For example, suppose $|0\rangle$ was measured, then the state of the system after this measurement is re-normalized as:

$$\frac{M_0 |0\rangle}{|a|} = \frac{a}{|a|} |0\rangle$$

Note that if the measurement is immediately repeated, then according to this rule the same outcome is attained again, with probability one.

3.6   Multi-qubit systems

The state space of a composite physical system is the tensor product of the state spaces of the component () physical systems. If the state of the component systems is $|\psi_i\rangle$, $i = \overline{1, n}$, then the joint state of the total system is

$$|\psi\rangle = |\psi_1\rangle \otimes |\psi_2\rangle \otimes \ldots \otimes |\psi_n\rangle = |\psi_1\rangle |\psi_2\rangle \ldots |\psi_n\rangle$$

where, $\otimes$ we denote the tensor product. Also, we can use the abbreviated notation of the tensor product $|\psi_1\rangle |\psi_2\rangle \ldots |\psi_n\rangle$.

❖ Example:

$$|\psi_1\rangle|\psi_2\rangle = \begin{bmatrix} 3 \\ 6i \end{bmatrix}\begin{bmatrix} 4 \\ 5 \end{bmatrix} = \begin{bmatrix} 3 \times \begin{bmatrix} 4 \\ 5 \end{bmatrix} \\ 6i \times \begin{bmatrix} 4 \\ 5 \end{bmatrix} \end{bmatrix} = \begin{bmatrix} 12 \\ 15 \\ 24i \\ 30i \end{bmatrix}$$

Suppose $|\psi_1\rangle = a|0\rangle + b|1\rangle$ and $|\psi_2\rangle = c|0\rangle + d|1\rangle$, then:

$$|\psi\rangle = |\psi_1\rangle \otimes |\psi_2\rangle = a \cdot c|0\rangle|0\rangle + a \cdot d|0\rangle|1\rangle + b \cdot c|1\rangle|0\rangle + b \cdot d|1\rangle|1\rangle$$

The quantum register for preceding 2 base quantum systems is

$$|\psi\rangle = \begin{pmatrix} a \cdot c \\ a \cdot d \\ b \cdot c \\ b \cdot d \end{pmatrix}$$

Tensor product works for classical systems (except the restricted definition of the probability amplitudes makes it so that the result is a simple concatenation). For quantum systems tensor product captures the essence of superposition, that is if system A is in state $|\psi\rangle_A$ and B in state $|\psi\rangle_B$ then there should be some way to have a little of A and a little of B.

3.7   Entanglement

Entanglement is a uniquely quantum phenomenon. Entanglement is one of the properties of quantum mechanics which caused Einstein and others to dislike the theory. In 1935, Einstein, Podolsky, and Rosen formulated the EPR paradox, demonstrating that entanglement makes quantum mechanics a non-local theory. Einstein famously derided entanglement as *spooky action at a distance*. Entanglement is a property of a multi-qubit state space (multi-qubit system) and can be thought of as a resource. To explain entanglement we will examine the creation and destruction of an EPR pair of qubits named after Einstein, Podolsky, and Rosen.

Consider two noninteracting systems A and B in the state $|\psi\rangle_A$ and $|\psi\rangle_B$ the state of the composite system is

$$(|\psi\rangle_A)(|\psi\rangle_B) = \left(\sum_i a_i |i\rangle_A\right)\left(\sum_j b_j |j\rangle_A\right)$$

States which can be represented in this form are called separable states. The most general state of, $|\psi\rangle_A \otimes |\psi\rangle_B$ which has the form

$$|\psi\rangle_A \otimes |\psi\rangle_B = \sum_{i,j} c_{i,j} |i\rangle_A |j\rangle_B$$

If such a state is not separable, it is known as an *entangled state*.

Suppose you begin with a qubit $|\psi_1\rangle$ in a zero $|0\rangle$ state. Let

$$U = H = \frac{1}{\sqrt{2}}\begin{bmatrix} 1 & 1 \\ 1 & -1 \end{bmatrix}$$

Then let $|\psi_1'\rangle = H|\psi_1\rangle = \frac{1}{\sqrt{2}}(|0\rangle + |1\rangle)$

Now take another qubit $|\psi_2\rangle$ also in the zero $|0\rangle$ state. The joint state-space probability vector is the tensor product of these two:

$$|\psi\rangle = |\psi_1'\rangle \otimes |\psi_2\rangle = \frac{1}{\sqrt{2}}|00\rangle + 0|01\rangle + \frac{1}{\sqrt{2}}|10\rangle + 0|11\rangle$$

Now define a new unitary transform:

$$CNot = \begin{bmatrix} 1 & 0 & 0 & 0 \\ 0 & 1 & 0 & 0 \\ 0 & 0 & 0 & 1 \\ 0 & 0 & 1 & 0 \end{bmatrix}$$

but for now lets just apply *CNot* to our two qubits:

$$|\psi'\rangle = CNot|\psi\rangle = \begin{bmatrix} 1 & 0 & 0 & 0 \\ 0 & 1 & 0 & 0 \\ 0 & 0 & 0 & 1 \\ 0 & 0 & 1 & 0 \end{bmatrix} \begin{bmatrix} 1/\sqrt{2} \\ 0 \\ 1/\sqrt{2} \\ 0 \end{bmatrix} = \frac{1}{\sqrt{2}} \left(|00\rangle + |11\rangle \right)$$

The key to entanglement is the property that the state space cannot be decomposed into component spaces. That is, there does not exists any $|\psi_1\rangle$ and $|\psi_2\rangle$ such that

$$|\psi_1\rangle \otimes |\psi_2\rangle = \frac{1}{\sqrt{2}} \left(|00\rangle + |11\rangle \right)$$

To illustrate why entanglement is so strange, let consider performing a measurement just prior to applying the *CNot* gate.

The two measurement operators M_{02} and M_{12} (for obtaining $|0\rangle$ or a $|1\rangle$) are:

$$M_{02} = \begin{bmatrix} 1 & 0 & 0 & 0 \\ 0 & 0 & 0 & 0 \\ 0 & 0 & 1 & 0 \\ 0 & 0 & 0 & 0 \end{bmatrix} \text{ and } M_{12} = \begin{bmatrix} 0 & 0 & 0 & 0 \\ 0 & 1 & 0 & 0 \\ 0 & 0 & 0 & 0 \\ 0 & 0 & 0 & 1 \end{bmatrix}$$

Recall that just prior to the *CNot* the system is in the state

44

$$|\psi\rangle = |\psi_1'\rangle \otimes |\psi_2\rangle = \frac{1}{\sqrt{2}}|00\rangle + 0|01\rangle + \frac{1}{\sqrt{2}}|10\rangle + 0|11\rangle$$

hence the result of measuring the second qubit will clearly be $|0\rangle$. Note that $M_{02}^{\dagger}M_{02} = M_{02}$. Therefore:

$$p(0) = \langle\psi|M_{02}|\psi\rangle = \begin{bmatrix} \frac{1}{\sqrt{2}} & 0 & \frac{1}{\sqrt{2}} & 0 \end{bmatrix} \begin{bmatrix} 1 & 0 & 0 & 0 \\ 0 & 0 & 0 & 0 \\ 0 & 0 & 1 & 0 \\ 0 & 0 & 0 & 0 \end{bmatrix} \begin{bmatrix} 1/\sqrt{2} \\ 0 \\ 1/\sqrt{2} \\ 0 \end{bmatrix} = 1$$

and after measurement:

$$\frac{M_m|\psi\rangle}{\sqrt{\langle\psi|M_m^{\dagger}M_m|\psi\rangle}} = \frac{|\psi\rangle}{1} = |\psi\rangle$$

We can see that measurement had no effect on the first qubit. It remains in a superposition of $|0\rangle$ and $|1\rangle$.

Now let consider the same measurement but just after the *CNot* gate is applied. Here:

$$|\psi'\rangle = CNot|\psi\rangle = \frac{1}{\sqrt{2}}(|00\rangle + |11\rangle)$$

Now it is not clear whether the second qubit will return $|0\rangle$ or $|1\rangle$, both outcomes are equally likely. To see this, let calculate the probability of obtaining $|0\rangle$:

$$p(0) = \langle \psi' | M_{02} | \psi' \rangle = \begin{bmatrix} \dfrac{1}{\sqrt{2}} & 0 & 0 & \dfrac{1}{\sqrt{2}} \end{bmatrix} \begin{bmatrix} 1 & 0 & 0 & 0 \\ 0 & 0 & 0 & 0 \\ 0 & 0 & 1 & 0 \\ 0 & 0 & 0 & 0 \end{bmatrix} \begin{bmatrix} 1/\sqrt{2} \\ 0 \\ 0 \\ 1/\sqrt{2} \end{bmatrix} = \dfrac{1}{2}$$

Hence, after the *CNot* gate is applied we have only a 1/2 chance of obtaining $|0\rangle$. Of particular interest is what happens to the state vector of the system after measurement:

$$\frac{M_m |\psi\rangle}{\sqrt{\langle \psi | M_m^\dagger M_m | \psi \rangle}} = \frac{M_{02} |\psi'\rangle}{\sqrt{1/2}} = \frac{1}{\sqrt{1/2}} \begin{bmatrix} \dfrac{1}{\sqrt{2}} \\ 0 \\ 0 \\ 0 \end{bmatrix} = \begin{bmatrix} 1 \\ 0 \\ 0 \\ 0 \end{bmatrix} = |00\rangle$$

This is the remarkable thing about entanglement. By measuring one qubit, we can affect the probability amplitudes of the other qubits of the system. How to think about this process in an abstract way is an open challenge in quantum computing. The difficulty is the lack of any classical analog.

One useful, but imprecise way to think about entanglement, superposition and measurement is quantum information. Entanglement links that information across quantum bits, but does not create any more of it. Measurement *destroys* quantum information turning it into classical. Thus, think of an EPR pair as having as much superposition as an unentangled set of qubits, one in a superposition between zero and one, and another in a pure state. The superposition in the EPR pair is simply linked across qubits instead of being isolated in one.

4 Quantum Computation in Matlab

The theory of computation began early in the twentieth century, before modern electronic computers had been invented. Computation can be defined as finding a solution to a problem from given inputs by means of an algorithm. For thousands of years, computing was done with pen and paper, or chalk and slate, or mentally, sometimes with the aid of tables.

Common present-day computers are built with hardware that is based primarily on the principles of classical physics. Most of these systems rely only on macroscopic physical properties. For example, an electric charge may be made of millions of electrons, or a bit on a hard disk may be comprised of many aligned iron particles.

The theory of quantum systems helps us predict the behavior of such systems as we decrease their scale towards the very small. As the Planck scale is approached, the behavior predicted for quantum systems is different from that expected by present-day hardware. Much work has been done to attempt to harness the behavior of quantum systems to create computers that are more efficient. A machine, which uses the properties of quantum systems to gain a performance advantage over other computers, has been called a *quantum computer*.

We have developed QCT (Quantum Computer Toolbox), a library written in Matlab, which simulates the capabilities of a *quantum computer*. Matlab is a well known in engineer academia as matrix computing environment, which makes it well suited for simulating quantum algorithms.

The QCT is written entirely in the Matlab and m-files are listed in current section. There are certain data types that are implicitly defined by the QCT, including data types for qubit registers and transformations. The QCT contains many functions designed to mimic the actions of a quantum computer. In addition, the QCT contains several convenience functions designed to aid in the creation and modification of the data types used in algorithms. It should be noted that Matlab starts counting matrix, list, vector, and array indices at 1.

Like many other classical simulations of quantum systems, QCT experiences an exponential slowdown and space blowup as the numbers of bits in the quantum register increases. Despite this, QCT can be used on moderate numbers of qubits. The QCT has been successfully used with quantum registers approaching eight bits in size. The main purposes of the QCT are for research involving *Quantum Computation* and as a teaching tool to aid in learning about Quantum Computing systems.

4.1   Introduction to Linear Algebra

In current subsection, we will see the capabilities of Matlab environment to implement typical computations used in quantum system theory. From point of view of Linear Algebra used in quantum systems theory the Matlab environment, ensure the best performance in matrix computation.

In Matlab environment is very easy to define and use complex numbers

    >> c = 2+3i

    c =

        2.0000 + 3.0000i

and complex conjugate can be easy obtained too

    >> c_conj = c'

    c_conj =

        2.0000 - 3.0000i

The ket vector $|\psi\rangle$ can be defined in Matlab environment as:

```
>> psi = [4; 5]

psi =

    4
    5
```

and bra vector $\langle\varphi| = [c_1^*, c_2^*, ..., c_n^*]$

```
>> phi = [3 -6i]

phi =

    3.0              0 - 6.0000i
```

The inner products $\langle\varphi|\psi\rangle$ between vectors $|\varphi\rangle$ and $|\psi\rangle$ is also very simple to compute in Matlab environment

```
>> psi = [4; 5]

psi =

    4
    5

>> phi = [3; 6i]

phi =

    3.0000
       0 + 6.0000i

>> phi_psi = phi'*psi

phi_psi =
```

12.0000 -30.0000i

The equality $\langle\varphi|\psi\rangle = \langle\psi|\varphi\rangle^*$ can be tested in few command line

```
>> psi_phi = psi'*phi

psi_phi =

  12.0000 +30.0000i

>> phi_psi = psi_phi'

phi_psi =

  12.0000 -30.0000i
```

The tensor product $|\varphi\rangle \otimes |\psi\rangle$ of two vectors $|\varphi\rangle$ and $|\psi\rangle$ use the *kron* function of the Matlab environment

```
>> psi = [4; 5]

psi =

   4
   5

>> phi = [3; 6i]

phi =

   3.0000
   0 + 6.0000i

>> phi_kron_psi = kron(phi,psi)

phi_kron_psi =

  12.0000
  15.0000
```

```
0 +24.0000i
0 +30.0000i
```

The computation of complex conjugate form of matrix A

```
>> A = [1 6i; 3i 2+4i]

A =

   1.0000          0 + 6.0000i
        0 + 3.0000i  2.0000 + 4.0000i
```

is identical with complex number case (in Matlab environment all variables are matrices)

```
>> A_conj = conj(A)

A_conj =

   1.0000          0 - 6.0000i
        0 - 3.0000i  2.0000 - 4.0000i
```

The Hermitian conjugate $A^\dagger = (A^T)^*$ (adjoint) of matrix A can be obtained by simple operations too

```
>> A = [1 6i; 3i 2+4i]

A =

   1.0000          0 + 6.0000i
        0 + 3.0000i  2.0000 + 4.0000i

>> A_adj = A'

A_adj =

   1.0000          0 - 3.0000i
        0 - 6.0000i  2.0000 - 4.0000i
```

To compute the norm

$$\left\| |\psi\rangle \right\| = \sqrt{\langle \psi | \psi \rangle}$$

of vector $|\psi\rangle$ in Matlab environment we can use two methods. First method is in according with definition in quantum systems theory

```
>> phi = [3; 6i]

phi =

   3.0000
   0 + 6.0000i

>> phi_norm = sqrt(phi'*phi)

phi_norm =

   6.7082
```

and a second method by Matlab function *norm*

```
>> phi = [3; 6i]

phi =

   3.0000
   0 + 6.0000i

>> phi_norm = norm(phi)

phi_norm =

   6.7082
```

The normalization of $|\psi\rangle$ defined by

$$|\psi\rangle/\||\psi\rangle\|$$

is a very important and frequently used procedure in quantum systems theory. We can implement this procedure very easy

```
>> phi = [3; 6i]

phi =

   3.0000
        0 + 6.0000i

>> phi_norm = sqrt(phi'*phi)

phi_norm =

   6.7082

>> norm_phi = phi/phi_norm

norm_phi =

   0.4472
        0 + 0.8944i

>> norm_norm_phi = norm(norm_phi)

norm_norm_phi =

   1
```

or using *norm* function from Matlab environment

```
>> norm_norm_phi = sqrt(norm_phi'*norm_phi)

norm_norm_phi =

   1.0000
```

The inner product $\langle\varphi|A|\psi\rangle$ of $|\varphi\rangle$ and $A|\psi\rangle$ or inner product of $A^\dagger|\varphi\rangle$ and $|\psi\rangle$ are given below

```
>> psi = [4; 5]

psi =

    4
    5

>> phi = [3; 6i]

phi =

   3.0000
        0 + 6.0000i

>> A = [1 6i; 3i 2+4i]

A =

   1.0000          0 + 6.0000i
        0 + 3.0000i   2.0000 + 4.0000i

>> phi_a_psi = phi'*A*psi

phi_a_psi =

   2.0400e+002 +3.0000e+001i

>> a_phi_psi = (A'*phi)'*psi

a_phi_psi =

   2.0400e+002 +3.0000e+001i
```

As we can see the Matlab environment can be perfect choice for quantum computation. The quantum algorithms can be easy write and tested without any additional special toolbox. However, few

Matlab function for frequently used procedure in quantum computation can decrease the programming and simulation effort. Therefore, a very good plan can be to create a folder ~/qct and generate here with Matlab editor the m-file from current section. This exercise can be very helpful to profound understand the basis of quantum computation and is necessary to create QCT functions.

```
contents.m

1    %   Qantum Computer Toolbox
2    %   ************************
3    %
4    %   (c) 2005 Ioan Burda
5    %
6    % functions:
7    %
8    %   mket        - Make KET
9    %   dket        - Dirac notation of the KET
10   %   nket        - Normalize KET
11   %   pket        - Plot KET
12   %   igm         - ImaGe of the Matrix
13   %   htm         - Hadamard Transform Matrix
14   %   qftm        - Quantum Fourier Transform Matrix
15   %   ufm         - Unitary Function Matrix
16   %   uim         - Unitary f-conditional Inverter Matrix
17   %   iam         - Inversion about Average Matrix
18   %   measure     - MEASURE
19
```

4.2   Make Ket

We represent an n-qubit ket register with a $2n-1$ Matlab matrix. The ith entry in each ket represents the complex amplitude of the ith basis element for example; the first element of a ket represents the amplitude of the basis element $|00...000\rangle$. There is a convenience function in QCT, *mket*, which will produce kets in standard basis.

```
mket.m

1      function psi = mket(state,n)
2      % mket  -  Make KET
3      %
4      % usage:
5      % psi = mket(state,n)
6      %                    n - qubits number
7
8      if n > 8
9         error('qubits number to big')
10     end
11     if state+1 > 2^n
12        error('state too big for register size')
13     end
14     %
15     psi = zeros(2^n,1);
16     psi(state+1,1) = 1;
17
```

We can test the QCT *mket* function from command line of the Matlab environment.

```
>> psi = mket(3,3)

psi =

     0
     0
     0
     1
     0
     0
     0
     0
```

4.3 Dirac Notation of the Ket

The Dirac notation of the ket is very useful and compact notation for state vector in quantum system. The listing of *dket* QCT functions is given below.

```
dket.m

1    function str_psi = dket(psi,arg)
2    % dket  -  Dirac notation of the KET
3    %
4    % usage:
5    % str_psi = dket(psi,arg)
6    %                      arg = ['bin']; 'dec'; 'hex'
7
8    if nargin == 1, arg = 'bin'; end
9    nb = log2(size(psi,1));
10   nn = 1 + fix((nb-1)/4);
11   str_psi = '';
12   for i = 1:length(psi)
13   if psi(i,1) ~= 0
14      if strcmp(arg,'bin')
15      str_psi = [str_psi,[num2str(psi(i,1)),...
16            '|',dec2bin((i-1),nb),'> + ']];
17      elseif strcmp(arg,'dec')
18      str_psi = [str_psi,[num2str(psi(i,1)),...
19            '|',num2str(i-1),'> + ']];
20      elseif strcmp(arg,'hex')
21      str_psi = [str_psi,[num2str(psi(i,1)),...
22            '|',dec2hex((i-1),nn),'> + ']];
23      end
24   end
25   end
26   str_psi = str_psi(1:end-3);
27
```

Next exercises are a good test for *dket* function.

```
>> psi=mket(3,3)

psi =

     0
     0
     0
     1
     0
     0
     0
     0

>> dket(psi)

ans =

1|011>

>> dket(psi,'dec')

ans =

1|3>

>> dket(mket(255,8),'hex')

ans =

1|FF>
```

4.4   Normalize Ket

There are also convenience functions to normalize a ket to unit
length *nket*. As we can see in the listing of *nket* QCT functions our
algorithm use *norm* functions of the Matlab environment.

nket.m	
1	`function psi = nket(psi)`
2	`% nket - Normalize KET`
3	`%`
4	`% usage:`
5	`% psi = nket(psi)`
6	
7	`psi = psi/norm(psi);`
8	

To produce a superposition of the basis elements, kets need only be added together with + from Matlab. This goal can be a very good exercise to test our *nket* QCT function.

```
>> psi1=mket(1,3)

psi1 =

     0
     1
     0
     0
     0
     0
     0
     0

>> psi2=mket(3,3)

psi2 =

     0
     0
     0
     1
     0
     0
     0
```

```
        0

>> psi = nket(psi+psi2)

psi =

        0
   0.7071
        0
   0.7071
        0
        0
        0
        0

>> dket(psi)

ans =

0.70711|001> + 0.70711|011>
```

4.5   Plot Ket

The QCT function *pket* show the probabilities of observing each basis state on a graph. This function is not necessary from point of view of the quantum systems theory but can be very useful to debug algorithms or to get a better illustration of some behavior of the algorithms.

The *stem* function of the Matlab environment is very easy to use and have high flexibility. For QCT purpose, the Matlab *stem* function is adapted to the specific request of quantum computation field. So, to generate this function in QCT we can use Matlab editor and listing given below.

```
pket.m

1    function pket(psi)
2    % pket  -  Plot KET
3    %
4    % usage:
5    % pket(psi)
6
7    if sum(imag(psi))==0
8        stem(0:length(psi)-1,real(psi).^2,'*r');
9    Ylabel('real part');
10   set(gca,'XLimMode','manual','XLim',[0 length(psi)-1],...
11     'YLimMode','manual','YLim',[0 1]);
12   else
13   %
14   subplot(2,1,1);
15       stem(0:length(psi)-1,real(psi).^2,'*r');
16   Ylabel('real part');
17   set(gca,'XLimMode','manual','XLim',[0 length(psi)-1],...
18     'YLimMode','manual','YLim',[0 1]);
19   %
20   subplot(2,1,2);
21       stem(0:length(psi)-1,imag(psi).^2,'*r');
22   Ylabel('imaginary part');
23   set(gca,'XLimMode','manual','XLim',[0 length(psi)-1],...
24     'YLimMode','manual','YLim',[0 1]);
25   %
26   set(0,'Units','pixels');
27   fsize=get(0,'ScreenSize');
28   set(gcf,'Units','pixels');
29   yfsize=fix((fsize(4)-64)/2) ;
30   xfsize= fix(yfsize*1.75) ;
31   set(gcf,'Position',...
32     [(fsize(3)-yfsize)/2 (fsize(4)-xfsize-48)/2 yfsize xfsize]);
33   %
34   end
35   drawnow;
36
```

\>> pket(psi)

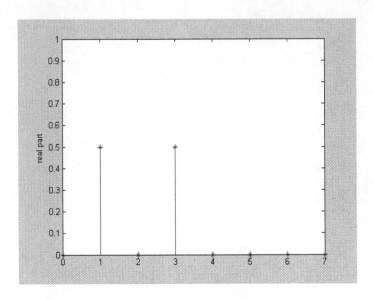

4.6 Image of the Matrix

The QCT function *igm* produce the image of the matrix and is useful in particular to debug algorithms. The listing of QCT function *igm* is given below.

igm.m	
1	function igm(um)
2	% igm - ImaGe of the Matrix
3	%
4	% usage:
5	% igm(um)
6	
7	if sum(sum(imag(um)))==0
8	[x y] = size(um);

```
9      surf(um);
10     view(2);
11     axis([1 x 1 y]);
12     Ylabel('real part');
13     set(gca,'YDir','reverse');
14     set(gca,'XAxisLocation','top');
15     colormap('cool');
16     else
17     %
18     subplot(2,1,1);
19     [x y] = size(um);
20     surf(real(um));
21     view(2);
22     axis([1 x 1 y]);
23     Ylabel('real part');
24     set(gca,'YDir','reverse');
25     set(gca,'XAxisLocation','top');
26     colormap('cool');
27     %
28     subplot(2,1,2);
29     [x y] = size(um);
30     surf(imag(um));
31     view(2);
32     axis([1 x 1 y]);
33     Ylabel('imaginary part');
34     set(gca,'YDir','reverse');
35     set(gca,'XAxisLocation','top');
36     colormap('cool');
37     %
38     set(0,'Units','pixels');
39     fsize=get(0,'ScreenSize');
40     set(gcf,'Units','pixels');
41     yfsize=fix((fsize(4)-64)/2);
42     xfsize= fix(yfsize*1.75);
43     set(gcf,'Position',...
44        [(fsize(3)-yfsize)/2 (fsize(4)-xfsize-48)/2 yfsize xfsize]);
45     %
46     end
47     drawnow;
48
```

4.7   Hadamard Transform Matrix

Transformations on an n-qubit register are represented as $2^n \times 2^n$ Matlab matrices, or in other words quantum gates. The user must guarantee that the matrix is in fact unitary, since checking that fact can be prohibitively expensive. To apply a transformation to a same-sized ket, we can simply use Matlab's matrix multiplication operator. The QCT function *htm* is used to create a general Hadamard Transform Matrix. The n-qubit Hadamard transformation takes a n-qubit ket in the basis state of $|000...00\rangle$ to an evenly weighted superposition of all n-qubit basis states. The *htm* function given in listing below use *hadamard* functions of the Matlab environment.

htm.m	
1	function m = htm(n)
2	% htm - Hadamard Transform Matrix
3	%
4	% usage:
5	% m = htm(n)
6	% n - qubits number
7	
8	if n > 8
9	error('qubits number is too big');
10	return
11	end
12	m = (1/sqrt(2^n))*hadamard(2^n);
13	

To test this function we can use direct command line facility or *igm* function of QCT.

```
>> htm(1)

ans =
```

```
0.7071   0.7071
0.7071  -0.7071
```

The QCT function *igm* is very useful to see the shape of big matrix.

```
>> igm(htm(5))
```

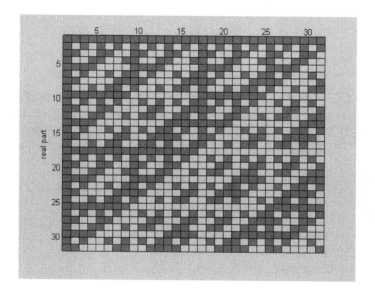

4.8   Quantum Fourier Transform Matrix

The Quantum Fourier Transform uses *normalized* basis functions (unlike the classical Discrete Fourier Transform) to represent a discrete state vector:

$$\langle j| = \frac{1}{\sqrt{2^n}} \sum_{j=0}^{2^n-1} \exp(2\pi i jk / 2^n) |k\rangle$$

As the basis is orthonormal, the Quantum Fourier Transform projections can be computed by the unitary transform:

$$mft = \frac{1}{\sqrt{2^n}} \begin{bmatrix} \omega^0 & \omega^1 & \omega^2 & \omega^3 & \dots \\ \omega^1 & \omega^2 & \omega^3 & \omega^4 & \dots \\ \cdot & \cdot & \cdot & \cdot & \dots \end{bmatrix}$$

where ω is the $(2^n)^{th}$ root of unity, $\exp(2\pi i / 2^n)$.

The QCT function *qftm* takes a single argument n, and produces the $n \times n$ Fourier transformation matrix. The listing of the *qftm* function is given below.

qftm.m
1   `function m = qftm(n)`

```
1    function m = qftm(n)
2    % qftm - Quantum Fourier Transformation Matrix
3    %
4    % usage:
5    % m = qftm(n)
6    %        n - qubits number, n <= 8
7    %
8    % m'*x   std -> f,  the same like FFT
9    % m *x   f -> std,  the same like IFFT
10
11   if n > 8
12      error('qubits number is too big');
13      return
14   end
15   w = exp(2*pi*i/(2^n));
16   m = zeros(2^n);
17   r = 0:2^n-1;
```

```
18   for i = 1:(2^n)
19       m(i,:)=r*(i-1);
20   end
21   m = w.^m;
22   m = m/sqrt(2^n);
23
```

The QCT provides a *qftm* function which creates Quantum Fourier Transform (QFT) matrices for any $n \le 8$. Note that the QFT is both unitary and hermetian, so the matrix obtained from the *qftm* function can be used to transform in both directions. In the following example, we test *qftm* function from QCT.

```
>> qftm(1)

ans =

    0.7071        0.7071
    0.7071       -0.7071 + 0.0000i

>> qftm(2)

ans =

Columns 1 through 3

    0.5000        0.5000              0.5000
    0.5000        0.0000 + 0.5000i   -0.5000 + 0.0000i
    0.5000       -0.5000 + 0.0000i    0.5000 - 0.0000i
    0.5000       -0.0000 - 0.5000i   -0.5000 + 0.0000i

Column 4

    0.5000
   -0.0000 - 0.5000i
   -0.5000 + 0.0000i
    0.0000 + 0.5000i
```

>> igm(qftm(5))

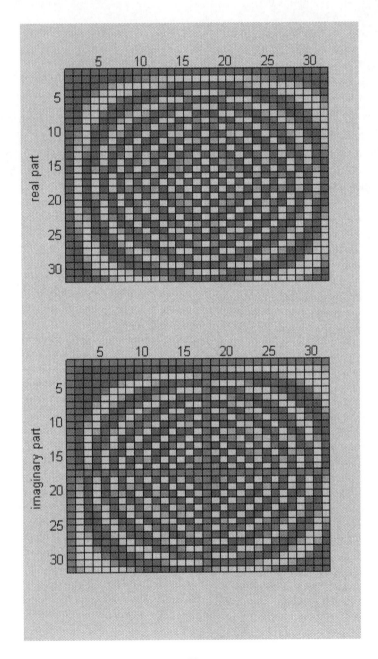

```
>> psi = mket(0,3)

psi =

     1
     0
     0
     0
     0
     0
     0
     0

>> psi = qftm(3)'*psi

psi =

     0.3536
     0.3536
     0.3536
     0.3536
     0.3536
     0.3536
     0.3536
     0.3536

>> psi = qftm(3)*psi

psi =

      1.0000
     -0.0000 + 0.0000i
     -0.0000 + 0.0000i
     -0.0000 + 0.0000i
          0 + 0.0000i
      0.0000 + 0.0000i
      0.0000 + 0.0000i
      0.0000 + 0.0000i
```

```
>> psi=mket(3,3)

psi =

    0
    0
    0
    1
    0
    0
    0
    0

>> psi=qftm(3)*psi

psi =

   0.3536
  -0.2500 + 0.2500i
  -0.0000 - 0.3536i
   0.2500 + 0.2500i
  -0.3536 + 0.0000i
   0.2500 - 0.2500i
   0.0000 + 0.3536i
  -0.2500 - 0.2500i
```

>> pket(psi)

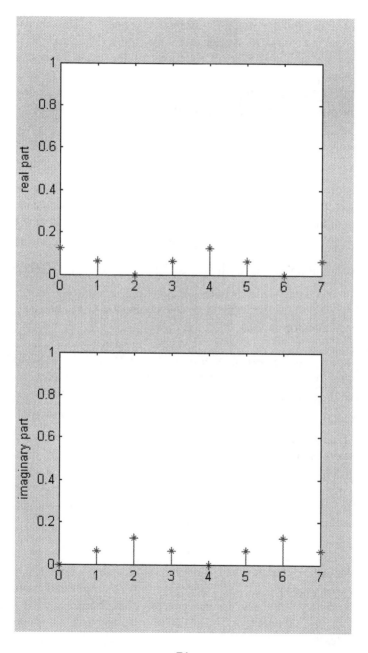

4.9 Unitary Function Matrix

More important for quantum computer concept is its *processing unit*, so we must to demonstrate the ability of this quantum processing unit to cover all functions of the classical computer processing unit. If we compare Boolean functions to unitary operators from a strictly functional point of view we can identify major differences between classical and quantum operations.

Since unitary operators, by definition, match the condition $U^\dagger U = I$, for every transformation U there exists the inverse transformation $U^\dagger$. As a consequence, quantum computation is restricted to reversible functions and an eigenstate can be transformed into a superposition of eigenstates. If the machine state $|\psi\rangle$ already is a superposition of several eigenstates, then a transformation $|\psi'\rangle = U|\psi\rangle$ is applied to all eigenstates simultaneously. This feature of quantum computing is called *quantum parallelism* and is a consequence of the linearity of unitary transformations.

In analogy to Boolean networks, unitary operators, which can be applied to arbitrary sets of qubits, are also referred to as *quantum gates*. A general Boolean function f can be implemented by using a pseudo-classical operator U. On the other hand, as a general n-qubit quantum state consists of maximally 2^n eigenstates with non-zero amplitude and unitary transformations take the form of linear operators and consequently can be described as

$$U : |i\rangle \to \sum_{j=0}^{2^n-1} u_{ij} |j\rangle \quad \text{with} \quad i, j \in Z^{2^n}$$

a classical computer can simulate any unitary operator with arbitrary precision by encoding the complex amplitudes as fixed point binary numbers. In the general case, however, this will require an exponential overhead in time as well as in space complexity.

One obvious problem of quantum computing is its restriction to reversible computations.

For any function $f : Z^{2^m} \rightarrow Z^{2^n}$ there exists a class of pseudo-classic operators $F : C^{2^{m+n}} \rightarrow C^{2^{m+n}}$ working on an m-qubits input and an n-qubits output register with $F|x,0\rangle = |x, f(x)\rangle$. Operators of that kind are referred to as quantum functions.

In other words, for any Boolean function $f : Z^{2^m} \rightarrow Z^{2^n}$ there exist $(2^{m+n} - 2^m)!$ different quantum functions F.

More generally, it is often useful to use unitary matrices to represent functions. In this representation, the functions can be performed in *parallel* on superposition of inputs. We define the matrix U_f by:

$$U_f(|i\rangle \otimes |0\rangle^{(n)}) = |x\rangle \otimes |f(i)\rangle \quad \text{with} \quad U_f^\dagger U_f = I$$

where there are m bits in the input, x, and n bits in the output. Note that the action of U_f is to replace the zero vector with the value of $f(x)$, whilst leaving the x in tact.

The action of U_f is to replace the zero vectors with the value of $f(x)$, whilst leaving the x in tact. So any classically computable function f can also be implemented on a quantum computer.

The QCT function *ufm* provides a quick way to construct such unitary matrices. To use it, we define an ordinary function $f(x,n)$, and pass it as an argument to the *ufm* constructor, along with specifications of the qubits number (n). The listing of the *ufm* QCT function is given below.

```
ufm.m

1       function um = ufm(f,m,n)
2       % ufm -- Unitary Function Matrix
3       %
4       % usage:
5       % um = ufm(f,m,n)
6       %               m - qubits in input
7       %                n - qubits in output
8       %
9
10      um = zeros(2^(m+n));
11      for i=0:2^m-1
12        for j=0:2^n-1
13          in = 2^n*i+j;
14          oj = 2^n*i + bitxor(j,feval(f,i,n));
15          um(oj+1,in+1)=1;
16        end
17      end
18
```

Here we define a simple function, *fmod*, which returns the modulus of $x+1$. We then display a visualization of the unitary $U_{f\,\text{mod}}$ (use a new folder ~/qct_e/qct_test) which represents it:

```
fmod.m

1       function md = fmod(x,n)
2       md=mod(x+1, 2^n);
3
```

```
>> igm(ufm('fmod',3,2))
```

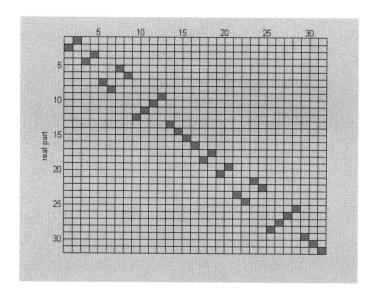

We now illustrate how the $U_{f \bmod}$ matrix can be used. We form the state $|\psi\rangle \otimes |0\rangle^{(2)}$ using the *kron* Matlab function for the tensor product. Applying $U_{f \bmod}$ to the result simultaneously computes the superposed results of *fmod* for the two superposed states comprising $|\psi\rangle$.

```
>> psi1=nket(mket(3,3)+mket(5,3))

psi1 =

       0
       0
       0
  0.7071
       0
  0.7071
       0
```

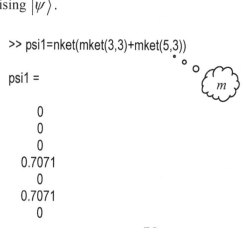

```
        0

>> dket(psi1)

ans =

0.70711|011> + 0.70711|101>

>> psi2=mket(0,2)

psi2 =

    1
    0
    0
    0

>> dket(psi2)

ans =

1|00>

>> psi = kron(psi1,psi2);
>> dket(psi)

ans =

0.70711|01100> + 0.70711|10100>

>> Ufmod=ufm('fmod',3,2);
>> psip=Ufmod*psi;
>> dket(psip) m

        ans =

        0.70711|01100> + 0.70711|10110>
```

A very interesting example can be obtained if we choice the case of function *isp* (~/qct_e/qct_test) which return value "1" if the input number is prime. (isprime(x) - Matlab function).

isp.m	
1 2 3	function is = isp(i,n) is = isprime(i);

First, we will prepare a superposition vector state $|\psi_1\rangle$.

```
>> psi1= mket(0,3)

psi1 =

    1
    0
    0
    0
    0
    0
    0
    0

>> psi1=qftm(3)*psi1

psi1 =

    0.3536
    0.3536
    0.3536
    0.3536
    0.3536
    0.3536
    0.3536
    0.3536
```

Also, the vector state $|\psi_2\rangle = |00...0\rangle$ is very easy to generate with *mket* QCT function.

```
>> psi2=mket(0,1)

psi2 =

    1
    0
```

Now, we can produce the $|\psi\rangle$ vector, which is the state of the whole system. Applying U_{isp} to the result simultaneously computes the superposed results of *isp* for the two superposed states comprising $|\psi\rangle$.

```
>> psi=kron(psi1,psi2);
>> dket(psi)

>> dket(psi)

ans =

0.35355|0000> + 0.35355|0010> + 0.35355|0100> +
0.35355|0110> + 0.35355|1000> + 0.35355|1010> +
0.35355|1100> + 0.35355|1110>

>> psip=Uisp*psi;
>> dket(psip)

ans =

0.35355|0000> + 0.35355|0010> + 0.35355|0101> +
0.35355|0111> + 0.35355|1000> + 0.35355|1011> +
0.35355|1100> + 0.35355|1111>
```

As we can see the last qubit is set "1" is the related number is prime number. The typical shape of U_{isp} is presented below.

```
>> igm(ufm('isp',4,1))
```

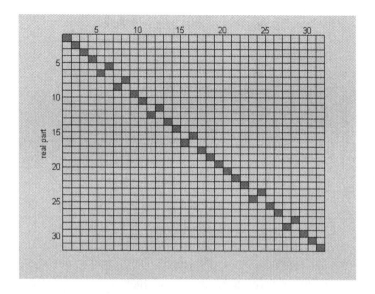

4.10 Unitary *f*-conditional Inverter Matrix

Classical programs allow the conditional execution of code in dependence on the content of a Boolean variable (conditional branching). A unitary operator, on the other hand, is static and has no internal flow control. Nevertheless, we can conditionally apply an n-qubit operator U to a quantum register by using an *enable* qubit and define a $n+1$ qubit operator

$$U' = \begin{bmatrix} I(n) & 0 \\ 0 & U \end{bmatrix}$$

So U is only applied to base-vectors where the enable bit is set. This can be easily extended to enable-registers of arbitrary length. Conditional operators a frequently used in arithmetic quantum functions and other pseudo-classic operators.

```
uim.m

1        function ui = uim(f,c,n)
2        % uim -- Unitary f-conditional Inverter Matrix
3        %
4        % usage:
5        % ui = uim(f,c,n)
6        %             c - condition
7        %             n - qbits number
8
9        ui = zeros(2^n);
10       for i=0:2^n-1
11           ui(i+1,i+1)=(-1)^(feval(f,c,i));
12       end
13
```

The QCT function *uim* presented here build unitary f-conditional inverter for n bit input such that:

$$U|\psi\rangle = (-1)^{f(\psi)}|\psi\rangle$$

In general, we can define a corresponding unitary matrix U_f, whose action on input strings $|\psi\rangle$ is to invert them if and only if $f|\psi\rangle = 1$ (Otherwise, it does nothing).

In particular case of *hay* function given below (use ~/qct_e/qct_test folder in Matlab work area to create it) we can see the action of U_{hay} matrix.

```
hay.m

1        function h = hay(cond,i)
2        h = 0;
3        if i == cond
4            h=1;
```

5	end
6	

```
>> Uhay = uim('hay',3,2)

Uhay =

    1    0    0    0
    0    1    0    0
    0    0    1    0
    0    0    0   -1

>> psi=htm(2)*mket(0,2)

psi =

    0.5000
    0.5000
    0.5000
    0.5000

>> psip = Uhay*psi

psip =

    0.5000
    0.5000
    0.5000
   -0.5000
```

The typical shape of U_{hay} for condition equal with seven is presented below.

```
>> igm(uim('hay',7,5))
```

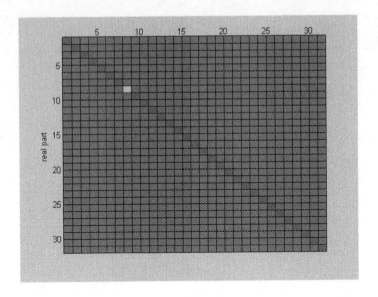

4.11 Inverse about Average

We define the *inversion about average* operation on our state vector as an operator that takes the amplitude of the ith state, and increases or decreases it. So that it is as much above or below the average as it was below or above the average before the operation.

The matrix representation of the inversion about average operator U_a is defined:

$$U_a = \begin{cases} 2/n, & if \ i \ne j \\ -1+2/n, & if \ i = j \end{cases}$$

Note that $U_a = -I + 2P$ where I is the identity matrix, and P is the matrix, which each element is equal to $1/n$. Observe that P has the following two properties, first $P^2 = P$, and second $P \cdot v$ for any v, results in a vector v' with each element being the arithmetic average of the element of v.

Now we can examine the operation of U_a on an arbitrary vector v

$$U_a \cdot v = (-I + 2P) \cdot v = -v + 2P \cdot v$$

By the second property of P above, note that $P \cdot v$ is a vector with each element equal to a where a is the arithmetic average of the elements of v. Therefore, the ith component of the vector is $(-v_i + 2a)$ which can be written $a + (a - v_i)$. Thus, the ith element is exactly as much above/below average as it was below/above average before the operation.

To show the inversion about average matrix U_a is unitary, recall that U_a can be written as $U_a = -I + 2P$ where I is the identity matrix and P is the matrix with each element is equal to $1/n$. Recall that $P^2 = P$. The matrix U_a is real and symmetric, so U_a is its own transposed complex conjugate, and we must show

$$U_a^2 = (-I + 2P)^2 = I^2 - 2P - 2P + 4P^2 = I - 4P + 4P = I$$

The QCT function *iam* with listing presented here create *inversion about average matrix* for n-qubit state vectors.

iam.m	
1	function m = iam(n)
2	% iam -- Inversion about Average Matrix
3	%
4	% usage:
5	% m = iam(n)
6	% n - qubits number
7	
8	m = 2*ones(2^n)/(2^n) - eye(2^n);

In the following example we test *iam* function from QCT.

```
>> Ua=iam(1)

Ua =

    0   1
    1   0
```

The typical shape of U_a for five qubits is presented below.

```
>> igm(iam(5))
```

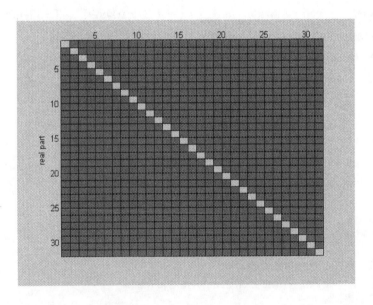

Additionally, an *inversion about the average matrix*, U_a, is used. The action of U_a on each element a_i of a state vector $|\psi\rangle$ is to replace it with $\bar{a}_j - a_i$.

```
>> Uhay=uim('hay',2,2)
```

84

```
Uhay =

    1    0    0    0
    0    1    0    0
    0    0   -1    0
    0    0    0    1

>> psi=htm(2)*mket(0,2)

psi =

    0.5000
    0.5000
    0.5000
    0.5000

>> psip=Uhay*psi

psip =

    0.5000
    0.5000
   -0.5000
    0.5000

>> psipp=Ua*psip

psipp =

    0
    0
    1
    0
```

4.12 Measure

Measuring a state with respect to the standard basis causes it to collapse into one of its standard basis eigenstates, with an eigenstate's probability given by the modulus of its squared amplitude. This non-deterministic process is simulated by the *measure* QCT function. The listing of measure QCT function is given below.

measure.m	
1	`function psi_o = measure(psi, qstr)`
2	`% measure  -  ket MEASURE`
3	`%`
4	`% usage:`
5	`%   psi_o = measure(psi, [qstr])`
6	`%              qstr - 'mx...',`
7	`%                     m - measure`
8	`%                     x - don't care`
9	`% length(qusrt) = n - qubits number`
10	
11	`n = log2(size(psi,1));`
12	`psi_o=zeros(size(psi));`
13	`psi=nket(psi);`
14	`%`
15	`if nargin == 1`
16	`    qstr(1:n)='m';`
17	`end`
18	`if nargin == 2 & length(qstr) ~= n`
19	`    error('length(qstr) ~= qubits number');`
20	`end`
21	`%`
22	`m = sum(qstr=='m');`
23	`x = sum(qstr=='x');`
24	`%`
25	`if m ==n`
26	`    [sp,ip]=sort(abs(psi).^2);`
27	`    sp=cumsum(sp);`

```
28      r = rand;
29      k=1;
30        while r>sp(k)
31          k=k+1;
32        end
33      psi_o(ip(k))=1;
34   else
35   %
36   ind=find(psi~=0);
37   for i=1:length(ind)
38     v(i).bin = dec2bin((ind(i)-1),n);
39     v(i).alpha = psi(ind(i));
40   end
41   %
42   [ms, mi] = sort(qstr);
43   mii(mi) = 1:length(mi);
44   for k=1:length(v)
45     v(k).bin = v(k).bin(mi);
46   end
47   %
48   cc = zeros(2^m,2^x);
49   %
50   for k=1:length(v)
51     ib = v(k).bin(1:m);
52     if isempty(ib)
53       error(['qstr = ',qstr]);
54     end
55     jb = v(k).bin(m+1:m+x);
56     i = bin2dec(ib);
57     j = bin2dec(jb);
58     cc(i+1,j+1) = v(k).alpha;
59   end
60   %
61   p_i = sum(abs(cc.^2), 2)';
62   psi_i = zeros(length(psi), 2^m);
63   %
64   for i=1:2^m
65   for j=1:2^x
66       if (cc(i,j)~=0)
67         bin = [dec2bin(i-1,m) dec2bin(j-1,x)];
```

87

```
68          bi = bin(mii);
69          psi_i(:,i) = psi_i(:,i) +...
70              cc(i,j) * mket(bin2dec(bi),length(bi));
71        end
72      end
73        if p_i(i)~=0
74            psi_i(:,i) = psi_i(:,i) / sqrt(p_i(i));
75        end
76      end
77      %
78      %simulate the collapse
79      %
80      [ps,pi]=sort(p_i);
81      sp=cumsum(ps);
82      r = rand;
83      k=1;
84      while r>sp(k)
85        k=k+1;
86      end
87      psi_o = psi_i(:,pi(k));
88      end
89
```

Here we prepare and measure a state. Note that the end line assigns the new state to the same variable *psi*, as the old state. This syntax should be used to simulate the real-world fact that the original information in $|\psi\rangle$ is irretrievably lost, and replaced by the collapsed $|\psi\rangle$:

```
>> psi =nket(mket(3,3)+mket(5,3))

psi =

     0
     0
     0
   0.7071
     0
```

88

```
        0.7071
             0
             0

>> dket(psi)

ans =

0.70711|011> + 0.70711|101>

>> psi = measure(psi)

psi =

             0
             0
             0
             0
             0
             1
             0
             0

>> dket(psi)

ans =

1|101>
```

In next example, we use *htm* and *pket* QCT function to create a more complex situation. As can be see after we create a superposition state with *htm* QCT function the effect of *measure* QCT function will be illustrated with *pket* QCT function. Observe that our example has the following two stages, first show the superposition state,

```
>> psi=mket(0,3) ;

>> psi=htm(3)*psi
```

```
psi =

    0.3536
    0.3536
    0.3536
    0.3536
    0.3536
    0.3536
    0.3536
    0.3536

>> pket(psi)
```

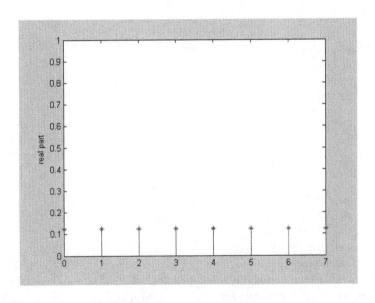

and second show the state vector after measurement as we can see below.

```
>> psi=measure(psi)

psi =
```

```
0
0
0
0
0
0
1
0
```

>> pket(psi)

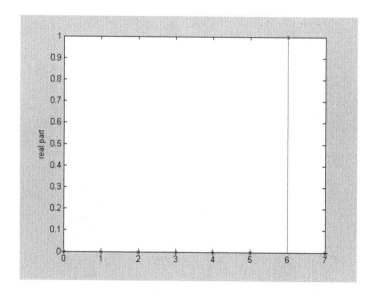

We consider the case of two (isolated) systems 1 and 2, with vector state described by the tensor product $|\psi\rangle = |\psi_1\rangle \otimes |\psi_2\rangle$. To an observer who has access only to system 1 or 2 the measured process is more difficult. A more general case is to make a measurement on only *some* of the qubits in a register of the quantum computer (the qubits who the observer has access). These results in the measured qubits collapsing to definite binary values, but the unmeasured qubits may still be left in a superposition. For example:

```
>> psi=nket(mket(9,6)+mket(26,6)+mket(40,6)+mket(52,6));
>> dket(psi)

ans =

0.5|001001> + 0.5|011010> + 0.5|101000> + 0.5|110100>

>> pket(psi)
```

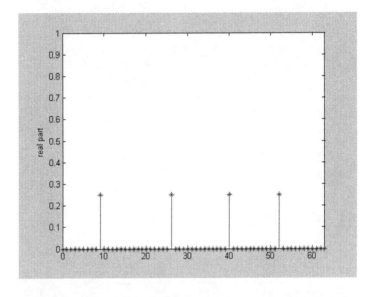

Suppose we measure qubits 2 and 5 (shown in teal bold above), leaving 1, 3, 4 and 6 unaffected. If the observed values of the measured subspace are 11 or 10 then we have collapsed into the second or fourth (respectively) states in the list above. However if the values are (0, 0) then we have collapsed into a superposition of the first and third states. The (1, 1) and (1, 0) outcomes each have a 1/4 chance of occurring, whilst the (0, 0) outcome has a 1/2 chance (due to it occurring in twice as many initial superposed states as the other outcomes). So to specify and perform the measurement from the above example, we can use:

```
>> psi=measure(psi,'xmxxmx');
```

Then look at the results in the next (random) example the state has collapsed to the most likely result, the (0, 0) case:

```
>> dket(psi)

ans =

0.70711|001001> + 0.70711|101000>

>> pket(psi)
```

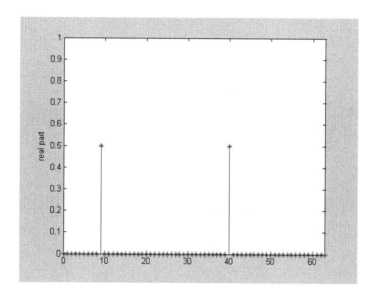

5 Quantum Algorithms

The Feynman's observation that classical systems cannot effectively model quantum mechanical systems generates interest in quantum computation. He proposed that the only way to effectively model a quantum mechanical system would be by using another quantum mechanical system (Feynman, 1982). Feynman's observation suggests that computers based on the laws of quantum mechanics instead of classical physics could be used to model quantum mechanical systems.

Deutsch was the first to explicitly ask whether it is possible to compute more efficiently on a quantum computer than on a classical computer. By addressing this question, he extended the theory of quantum computation further with the development of the universal quantum computer and quantum Turing machine (Deutsch, 1985) and with quantum computational networks (Deutsch, 1989). Deutsch also devised the first quantum algorithm, Deutsch's two bit problem (Deutsch 1985), this problem can be generalised to Deutsch's algorithm for finding out whether a function is balanced or constant (Deutsch & Jozsa, 1992).

Until the mid 1990's, quantum computation remained a curiosity. Although various uses had been suggested for quantum computers and some theory had been established. This situation changed when Shor published his quantum factoring algorithm for finding the prime factors of large integers (Shor, 1994). Finding prime factors is the basis of many public key encryption systems such as RSA and subsequently Shor's algorithm caused much interest within many sections of the scientific community.

Other algorithms such as that for discrete logarithms (Shor, 1994), an alternative factoring algorithm (Jozsa, 1997) based on Kitaev's work on construction of quantum algorithms based on group-theoretic principles (Kitaev, 1995), algorithms for finding the median and mean (Grover, 1997), Hogg's constraint satisfaction algorithms (Hogg, 1996) and Grover's algorithm for database search (Grover, 1997) all contribute to the relatively small number of known quantum algorithms.

Known quantum algorithms can be partitioned into three groups depending on the methods they use. The first group contains algorithms which are based on determining a common property of all the output values such as the period of a function, e.g. Shor's algorithm, the second contains those which transform the state to increase the likelihood that the output of interest will be read (amplification), e.g. Grover's algorithm and the third contains algorithms which are based on a combination of methods from the previous two groups. It is not known whether additional types of quantum algorithm exist or whether every quantum algorithm can be classified as a member of one of a finite number of groups.

Quantum computation simulation languages and systems have been developed which attempt to allow simulations of quantum algorithms, these include QCL (Ömer, 1998), Q-gol (Baker, 1997), Qubiter (Tucci, 1998) and the simulation system currently being developed by the OpenQubit group (OpenQubit, 1998). Q-gol was an attempt to write a high level programming language to allow researchers to describe algorithms designed to run on quantum computers. Qubiter takes as input an arbitrary unitary matrix and returns as output an equivalent sequence of elementary operations. Together with simulations produced within mathematical toolkits and implementation of algorithms using actual qubits, this has allowed verification that the known quantum algorithms work and enabled investigation into how they function.

As development work has progressed, additional uses have been proposed for quantum computation, from modelling quantum mechanical systems, breaking public key encryption, searching databases, generating true random numbers to providing secure communication using quantum key distribution. It has also been suggested that quantum mechanics may be playing a role in consciousness, if a quantum mechanical model of mind and consciousness was developed this would have significant impact on computational and artificial intelligence. At present, the majority of the research effort in quantum computation is devoted to the physics orientated aspects of quantum computation, in particular the development of hardware.

5.1   Quantum Teleportation

Quantum teleportation is an algorithm by which the complete quantum state of one qubit is transferred to another, while destroying the state of the original qubit. In other words, quantum teleportation is a means to replace the *state* of one qubit with that of another. It gets its out-of-this-world name from the fact that the state is *transmitted* by setting up an entangled state-space of three qubits and then removing two qubits from the entanglement (via measurement). Since the information of the source qubit is preserved by these measurements that *information* (i.e. state) ends up in the final third, destination qubit. This occurs, however, without the source (first) and destination (third) qubit ever directly interacting. The interaction occurs via entanglement. The algorithm to do this is also fairly simple, suppose:

$$|\psi_1\rangle = a \cdot |0\rangle + b|1\rangle$$

and given an EPR pair

$$|\psi_2\rangle = \frac{1}{\sqrt{2}}(|00\rangle + |11\rangle)$$

the state of the entire system is:

$$|\psi\rangle = |\psi_1\rangle \otimes |\psi_2\rangle = \frac{1}{\sqrt{2}}[a \ 0 \ 0 \ a \ b \ 0 \ 0 \ b]^T$$

❖   Example:

```
>> psi1=nket(mket(0,1)+mket(1,1))

psi1 =
```

```
    0.7071
    0.7071

>> dket(psi1)

ans =

0.70711|0> + 0.70711|1>

>> psi2=nket(mket(0,2)+mket(3,2))

psi2 =

    0.7071
         0
         0
    0.7071

>> dket(psi2)

ans =

0.70711|00> + 0.70711|11>

>> psi=kron(psi1,psi2)

psi =

    0.5000
         0
         0
    0.5000
    0.5000
         0
         0
    0.5000

>> dket(psi)

ans =
```

0.5|000> + 0.5|011> + 0.5|100> + 0.5|111>

Perform the *CNot* operation and we obtain

$$|\psi_2'\rangle = CNot \cdot |\psi\rangle = \frac{1}{\sqrt{2}}(|10\rangle + |01\rangle)$$

$$|\psi'\rangle = a \cdot |0\rangle \otimes |\psi_2\rangle + b \cdot |1\rangle \otimes |\psi_2'\rangle = \frac{1}{\sqrt{2}}[a \ 0 \ 0 \ a \ 0 \ b \ b \ 0]^T$$

First we create *cnot* function in ~/qct_e/qct_test folder. The listing of the *cnot* function is given below.

cnot.m	
1	function m = cnop
2	m = [0 1 0 0; 1 0 0 0; 0 0 0 1; 0 0 1 0];
3	

```
>> psi2p=cnop*psi2

psi2p =

        0
   0.7071
   0.7071
        0

>> psip=nket(kron(mket(0,1),psi2)+kron(mket(1,1),psi2p))

psip =

   0.5000
        0
        0
   0.5000
```

```
        0
   0.5000
   0.5000
        0
```

Next, we apply the *hadamard* gate (*htm* QCT function). However, as an aside, let examine what happens when we apply the *hadamard* gate to $|0\rangle$ and to $|1\rangle$. Recall that:

$$H = \frac{1}{\sqrt{2}}\begin{bmatrix} 1 & 1 \\ 1 & -1 \end{bmatrix}$$

$$H|0\rangle = \frac{1}{\sqrt{2}}\begin{bmatrix} 1 & 1 \\ 1 & -1 \end{bmatrix}\begin{bmatrix} 1 \\ 0 \end{bmatrix} = \frac{1}{\sqrt{2}}\begin{bmatrix} 1 \\ 1 \end{bmatrix}$$

$$H|1\rangle = \frac{1}{\sqrt{2}}\begin{bmatrix} 1 & 1 \\ 1 & -1 \end{bmatrix}\begin{bmatrix} 0 \\ 1 \end{bmatrix} = \frac{1}{\sqrt{2}}\begin{bmatrix} 1 \\ -1 \end{bmatrix}$$

Thus, applying H to our system we have:

$$|\varphi\rangle = a \cdot H|0\rangle \otimes |\psi_2\rangle + b \cdot H|1\rangle \otimes |\psi_2'\rangle =$$
$$\frac{1}{\sqrt{2}}\begin{bmatrix} a & b & b & a & a & -b & -b & a \end{bmatrix}^T$$

❖ Example:

```
>> phi=nket(kron(htm(1)*mket(0,1),psi2)+kron(htm(1)*mket(1,1),psi2p))

    phi =

    0.3536
    0.3536
    0.3536
    0.3536
    0.3536
   -0.3536
```

100

-0.3536
0.3536

We can rewrite this expression as:

$$|\varphi\rangle = \frac{1}{2}[|00\rangle(a\cdot|0\rangle + b\cdot|1\rangle) + |01\rangle(a|1\rangle + b\cdot|0\rangle) +$$
$$|10\rangle(a\cdot|0\rangle - b\cdot|1\rangle) + |11\rangle(a\cdot|1\rangle - b\cdot|0\rangle)]$$

Which we can shorten to:

$$|\varphi\rangle = \frac{1}{2}[|00\rangle\begin{bmatrix} 1 & 0 \\ 0 & 1 \end{bmatrix}|\psi_1\rangle + |01\rangle\begin{bmatrix} 0 & 1 \\ 1 & 0 \end{bmatrix}|\psi_1\rangle +$$
$$|10\rangle\begin{bmatrix} 1 & 0 \\ 0 & -1 \end{bmatrix}|\psi_1\rangle + |11\rangle i\begin{bmatrix} 0 & -i \\ i & 0 \end{bmatrix}|\psi_1\rangle]$$

These gates are the famous Pauli I, X, Z, Y operators and this is also written as:

$$|\varphi\rangle = \frac{1}{2}[|00\rangle I|\psi_1\rangle + |01\rangle X|\psi_1\rangle + |10\rangle Z|\psi_1\rangle + |11\rangle iY|\psi_1\rangle]$$

And of interest to us with teleportation:

$$|\varphi\rangle = \frac{1}{2}[|00\rangle I|\psi_1\rangle + |01\rangle X|\psi_1\rangle + |10\rangle Z|\psi_1\rangle + |11\rangle XZ|\psi_1\rangle]$$

This implies that we can measure the first and second qubit and obtain two classical bits. These two classical bits tell us what transform was applied to the third qubit.

Thereby we can *fixup* the third qubit by knowing the classical outcome of the measurement of the first two qubits. This fixup   is fairly straightforward, either applying nothing, X, Z or both X or Z. Lets work through an example:

101

$$M_{10} = \begin{bmatrix} 0 & 0 & 0 & 0 & 0 & 0 & 0 & 0 \\ 0 & 0 & 0 & 0 & 0 & 0 & 0 & 0 \\ 0 & 0 & 0 & 0 & 0 & 0 & 0 & 0 \\ 0 & 0 & 0 & 0 & 0 & 0 & 0 & 0 \\ 0 & 0 & 0 & 0 & 1 & 0 & 0 & 0 \\ 0 & 0 & 0 & 0 & 0 & 1 & 0 & 0 \\ 0 & 0 & 0 & 0 & 0 & 0 & 0 & 0 \\ 0 & 0 & 0 & 0 & 0 & 0 & 0 & 0 \end{bmatrix}$$

$P(10) = \langle \varphi | M_{10}^{\dagger} M_{10} | \varphi \rangle = \langle \varphi | M_{10} | \varphi \rangle$, since here $M_{10}^{\dagger} M_{10}$. Thus:

$$M_{10} | \varphi \rangle = \frac{1}{2} \begin{bmatrix} 0 & 0 & 0 & 0 & a & -b & 0 & 0 \end{bmatrix}^{T}$$

Recall that by definition of a qubit we know that $a \cdot a^{*} + b \cdot b^{*} = 1$, hence the probability of measuring 01 is 1/4. The same is true for the other outcomes.

❖ Example:

```
>> phi=measure(phi,'mmx')

phi =

       0
       0
       0
       0
  0.7071
 -0.7071
       0
       0
```

102

5.2 Deutsch's Algorithm

David Deutsch in 1985 emphasized that a quantum computer can best realize its computational potential by invoking what he called *quantum parallelism*. To understand what this means, it is best to consider an example. Following Deutsch, imagine we have a black box that computes a function that takes a single bit x to a single bit $f(x)$. We do not know what is happening inside the box, but it must be something complicated, because the *computation time* t_c is very long.

There are four possible functions $f(x)$ (because each of $f(0)$ and $f(1)$ can take either one of two possible values) and we would like to know what the box is computing. It would take $2t_c$ to find out both $f(0)$ and $f(1)$. But we need the answer in computation time t_c, not $2t_c$. Moreover, it turns out that we would be satisfied to know whether $f(x)$ is constant $f(0) = f(1)$ or balanced $f(0) \neq f(1)$.

Now suppose we have a quantum black box that computes $f(x)$. Of course $f(x)$ might not be invertible, while the action of our quantum computer is unitary and must be invertible, so we will need a transformation U_f that takes two qubits to two:

$$U_f : |x\rangle|y\rangle \rightarrow |x\rangle|y \oplus f(x)\rangle$$

where $\oplus$ represents addition modular two. This machine flips the second qubit if $f(x)$ acting on the first qubit is 1, and doesn't do anything if $f(x)$ acting on the first qubit is 0. We can determine if $f(x)$ is constant or balanced by using the quantum black box twice. However, it still takes a day for it to produce one output, so that will not do. We can get the answer by running the quantum black box just once. Because the black box is a quantum computer, we can choose the input state to be a superposition of $|0\rangle$ and $|1\rangle$. If the second qubit is initially prepared in the state $1/\sqrt{2}\,(|0\rangle+|1\rangle)$, then

$$U_f : |x\rangle \frac{1}{\sqrt{2}}(|0\rangle+|1\rangle) \to |x\rangle \frac{1}{\sqrt{2}}(|f(x)\rangle - |1\oplus f(x)\rangle) =$$

$$|x\rangle(-1)^{f(x)}\frac{1}{\sqrt{2}}(|0\rangle-|1\rangle)$$

so we have isolated the function $f(x)$ in an x-dependent phase. Now suppose we prepare the first qubit as $1/\sqrt{2}\,(|0\rangle+|1\rangle)$. Then the black box acts as

$$U_f : \frac{1}{\sqrt{2}}(|0\rangle+|1\rangle)\frac{1}{\sqrt{2}}(|0\rangle-|1\rangle) \to$$

$$\frac{1}{\sqrt{2}}[(-1)^{f(0)}|0\rangle+(-1)^{f(1)}|1\rangle]\frac{1}{\sqrt{2}}(|0\rangle-|1\rangle)$$

Finally, we can perform a measurement that projects the first qubit onto the basis

$$|\pm\rangle = \frac{1}{\sqrt{2}}(|0\rangle\pm|1\rangle)$$

Evidently, we will always obtain $|+\rangle$ if the function is balanced, and $|-\rangle$ if the function is constant. Therefore, we have solved Deutsch's problem, and we have found a separation between what a classical computer and a quantum computer can achieve. The classical computer has to run the black box twice to distinguish a balanced function from a constant function, but a quantum computer does the job in one run. This is possible because the quantum computer is not limited to computing either $f(0)$ or $f(1)$. It can act on a superposition of $|0\rangle$ and $|1\rangle$, and thereby extract *global* information about the function, information that depends on both $f(0)$ and $f(1)$. This is quantum parallelism.

Deutsch's was one of the first quantum algorithms to demonstrate an effect unobtainable by classical computation. Suppose we have four functions: dc0, dc1, db0 and db1 (~/qct_e/deutsch) as given below

dc0.m

```
1    function y = dc0(x,n)
2    y = 0;
3
```

dc1.m

```
1    function y = dc1(x,n)
2    y = 1;
3
```

db0.m

```
1    function y = db0(x,n)
2    y = x;
3
```

db1.m

```
1    function y = db1(x,n)
2    y = ~x;
3
```

The dc* functions are said to be *constant* and the db* functions are *balanced*. Given a black box, which computes U_f, where $f(x)$ a function is chosen from the above set, the task is to determine whether the function represented by U_f is constant or balanced.

A classical machine would require two evaluations of the black box. However, Deutsch's algorithm provides a way to do it in just one evaluation, on a quantum computer. Here is the Matlab code (~/qct_e/deutsch) that implements Deutcsh's Algorithm.

deutsch.m

```
1    function deutsch(f, arg)
2    % deutsch  -- DEUTSCH's algorithm
3    %
4    % written by Ioan Burda, (c) 2005
5
6    if ~nargin
7        error('usage: deutch(fname,[htm/ufm]')
8    end
9    if (nargin == 1)
10       arg = 'none';
11   end
12   %
13   psi = mket(1,2);
14   disp(['|psi> = ', dket(psi)]);
15   disp('... unitary function matrix');
16   Uf = ufm(f, 1, 1);
17   if strcmp(arg,'ufm')
18       disp(num2str(Uf));
19   end
20   disp('... hadamard transform matrix');
21   H = htm(2);
22   if strcmp(arg,'htm')
23       disp(num2str(H));
24   end
25   disp(['... H*Uf*H*|psi>']);
26   psi = H*Uf*H*psi;
27   disp('... measure |psi>');
```

28	psi = measure(psi);
29	disp(['|psi> = ', dket(psi)]);
30	

Note the *deutsch* function has been defined so that a test function is passed in as an argument. The results show the values of *psi* at the end of the algorithm, for the four black-box functions. Looking at the first qubit of the final states tells us whether the function is constant (0) or balanced (1).

```
>> deutsch('dc0')
|psi> = 1|01>
... unitary function matrix
... hadamard transform matrix
... H*Uf*H*|psi>
... measure |psi>
|psi> = 1|01>

>> deutsch('dc1')
|psi> = 1|01>
... unitary function matrix
... hadamard transform matrix
... H*Uf*H*|psi>
... measure |psi>
|psi> = 1|01>

>> deutsch('db0')
|psi> = 1|01>
... unitary function matrix
... hadamard transform matrix
... H*Uf*H*|psi>
... measure |psi>
|psi> = 1|11>

>> deutsch('db1')
|psi> = 1|01>
... unitary function matrix
... hadamard transform matrix
... H*Uf*H*|psi>
```

```
... measure |psi>
|psi> = 1|11>
```

Deutsch's algorithm was one of the first algorithms to appear for a quantum device. In next two examples, we can see the hadamard and unitary function matrix. The unitary function matrix is responsible for system state evolution.

```
>> deutsch('db1','ufm')
|psi> = 1|01>
... unitary function matrix
0 1 0 0
1 0 0 0
0 0 1 0
0 0 0 1
... hadamard transform matrix
... H*Uf*H*|psi>
... measure |psi>
|psi> = 1|11>

>> deutsch('db1','htm')
|psi> = 1|01>
... unitary function matrix
... hadamard transform matrix
0.5     0.5     0.5     0.5
0.5     -0.5    0.5     -0.5
0.5     0.5     -0.5    -0.5
0.5     -0.5    -0.5    0.5
... H*Uf*H*|psi>
... measure |psi>
|psi> = 1|11>
```

5.3   Deutsch - Jozsa Algorithm

The Deutsch-Jozsa algorithm is a quantum algorithm, proposed by David Deutsch and Richard Jozsa in 1992. The Deutsch-Jozsa algorithm is a generalization of Deutsch's algorithm. It was one of the first examples of a quantum algorithm, which is a class of algorithms designed for execution on quantum computers and have the potential to be more efficient than conventional, classical, algorithms by taking advantage of the quantum superposition and entanglement principles.

Suppose $f(x):\{2^{n-1}\} \rightarrow \{0,1\}$ and that $f(x)$ is either constant or balanced. The goal is determine which one it is. Classically it is trivial to see that this would require (in worst case) querying just over half the solution space, or $2^{n-1}/2+1$ queries. The Deutsch-Jozsa algorithm answers this question with just *one* query.

In Deutsch-Jozsa problem, we are given a black box computing a 0-1 valued function $f(x)$. The black box takes n bits and outputs the value $f(x)$. We know that the function in the black box is either constant (0 on all inputs or 1 on all inputs) or *balanced* (returns 1 for half the domain and 0 for the other half). The task is to determine whether $f(x)$ is constant or balanced.

For a conventional deterministic algorithm, 2^{n-1} evaluations of $f(x)$ will be required in the worst case. For a conventional randomized algorithm, a constant number of evaluation suffices to produce the correct answer with a high probability but 2^{n-1} evaluations are still required if we want an answer that is always correct. The Deutsch-Jozsa quantum algorithm produces an answer that is always correct with just one evaluation of $f(x)$.

The algorithm is as follows. First, do Hadamard transformations (H) on n 0s, forming all possible inputs, and a single 1, which will be the answer qubit. Next, run the function once; this XORs the result with the answer qubit. Finally, do Hadamards on the n inputs again, and measure the answer qubit. If it is 0, the function is constant, otherwise the function is balanced.

The starting state of the system $|\psi_0\rangle$ is fairly straightforward

$$|\psi_0\rangle = |0\rangle^{\otimes n}|1\rangle$$

The symbolic notation $|0\rangle^{\otimes n}$ simply means n consecutive $|0\rangle$ qubits. We then apply the $H^{\otimes n}$ transform. This symbol means to apply the H gate to each of the n qubits, in parallel. One way to define this transform is:

$$H^{\otimes n}|i\rangle = \sum_j \frac{(-1)^{i \cdot j}}{\sqrt{2^n}}|j\rangle$$

This notation is rather terse, but what it is saying is that given any arbitrary state vector, it will be composed of components $|i\rangle$. Each component of this state vector is transformed into a superposition of components $|j\rangle$. For example, let examine a single qubit:

$$a|0\rangle + b|1\rangle$$

Apply $H^{\otimes 1}$ to get:

$$a\frac{(-1)^{0 \cdot 0}}{\sqrt{2}}|0\rangle + a\frac{(-1)^{0 \cdot 1}}{\sqrt{2}} + b\frac{(-1)^{1 \cdot 0}}{\sqrt{2}} + b\frac{(-1)^{1 \cdot 1}}{\sqrt{2}} =$$

$$\frac{1}{\sqrt{2}}[(a+b)|0\rangle + (a-b)|1\rangle]$$

When we look at the actual transform as we have been writing it in the past we find:

$$\frac{1}{\sqrt{2}}\begin{bmatrix}1 & 1 \\ 1 & -1\end{bmatrix}\begin{bmatrix}a \\ b\end{bmatrix} = \frac{a+b}{\sqrt{2}}|0\rangle + \frac{a-b}{\sqrt{2}}|1\rangle$$

Returning to starting state of the system, $|\psi_0\rangle$ we transform it by:

$$|\psi_1\rangle = H^{\otimes n}|0\rangle^{\otimes n} H|1\rangle = \sum_{x \in \{0,1\}^n} \frac{1}{\sqrt{2}}|x\rangle\left[\frac{|0\rangle - |1\rangle}{\sqrt{2}}\right]$$

The notation $\{0,1\}^n$ means all possible bit strings of size n. We then apply the transform U_f that implements $f(x)$ to obtain the state $|\psi_2\rangle$:

$$|\psi_2\rangle = \sum_{x \in \{0,1\}^n} \frac{(-1)^{f(x)}}{\sqrt{2^n}}|x\rangle\left[\frac{|0\rangle - |1\rangle}{\sqrt{2}}\right]$$

Finally we apply another $H^{\otimes n}$ transform to obtain $|\psi_3\rangle$:

$$|\psi_3\rangle = \sum_{z \in \{0,1\}^n} \sum_{x \in \{0,1\}^n} \frac{(-1)^{x \cdot z + f(x)}}{\sqrt{2^n}}|z\rangle\left[\frac{|0\rangle - |1\rangle}{\sqrt{2}}\right]$$

The key to the Deutsch-Jozsa algorithm is the rather subtle point: Observe the probability amplitude of $z = |0\rangle^{\otimes n}$. Consider when $f(x)$ is constant. Since $z = |0\rangle^{\otimes n}$, we know that $(-1)^{x \cdot z + f(x)}$ is either -1 or $+1$ for all values of x. Hence, if $f(x)$ is constant the probability amplitude for $z = |0\rangle^{\otimes n}$ is expressed as:

$$\sum_{x \in \{0,1\}^n} \frac{1}{2^n} = \pm 1$$

111

Hence, when you measure the query register you will obtain a zero. Since postulate one tells that the probabilities must sum to 1, if $f(x)$ is constant, then we must measure a zero.

On the other hand, let consider if $f(x)$ is balanced. Then $(-1)^{x \cdot z + f(x)}$ will be $+1$ for some x and -1 for other x's. This is where the balanced requirement comes into play. Since all x's are considered, and the function is perfectly balanced, the probability of obtaining $z = |0\rangle^{\otimes n}$ is expressed as:

$$\sum_{x_1} \frac{+1}{2^n} + \sum_{x_2} \frac{-1}{2^n} = 0$$

Where x_1 is the set of x's such that $f(x)$ is equal to 0 and x_2 are those x's where $f(x)$ is equal to 1.

As with the original Deutsch algorithm, the Deutsch-Jozsa problem can be solved with just one black box evaluation of the U_f. This quantum computation exhibits *massive quantum parallelism*. Suppose we have four functions: djc0, djc1, djb0 and djb1 (~qct_e/deutsch_j) as given below

djc0.m	
1	function y = djc0(x,n)
2	y = 0;
3	

djc1.m	
1	function y = djc1(x,n)
2	y = 1;
3	

djb0.m

```
1    function y = djb0(x,n)
2    y = mod(x,2);
3
```

djb1.m

```
1    function y = djb1(x,n)
2    y = ~mod(x,2);
3
```

Matlab/QCT code (~/qct_e/deutsch_j) for Deutsch-Jozsa and some example test functions are shown below.

deutsch_jozsa.m

```
1    function deutsch_jozsa(f,arg)
2    % deutsch_jozsa -- Deutsch-Jozsa algorithm
3    %
4    % written by Ioan Burda, (c) 2005
5
6    if ~nargin
7       error('usage: deutch_jozsa(fname,[htm/ufm]')
8    end
9    if (nargin == 1)
10      arg = 'none';
11   end
12   if (nargin == 2)
13      h = figure(1);
14      set(h,'MenuBar','none','NumberTitle','off');
15   if strcmp(arg,'ufm')
16      set(h,'Name','Unitary Function Matrix');
17   elseif strcmp(arg,'htm')
18      set(h,'Name','Hadamard Transform Matrix');
19   end
```

113

```
20    end
21    %
22    psi = mket(1,6);
23    disp(['|psi> = ', dket(psi)]);
24    %
25    disp('... unitary function matrix');
26    Uf = ufm(f, 6-1, 1);
27    if strcmp(arg,'ufm')
28       igm(Uf);
29    end
30    %
31    disp('... hadamard transform matrix');
32    H6 = htm(6);
33    H6i = kron(htm(6-1),eye(2^1));
34    if strcmp(arg,'htm')
35       igm(H6);
36    end
37    %
38    disp(['... H*Uf*H*|psi>']);
39    psi = H6i*Uf*H6*psi;
40    %
41    disp('... measure |psi>');
42    psi = measure(psi);
43    disp(['|psi> = ', dket(psi)])
44
```

Note the *deutsch_josza* function has been defined so that a test function is passed in as an argument. The final, measured, *psi* state is interpreted by looking at the first $n-1$ qubits. If they are all zero then the function is constant. Otherwise, it is balanced. The images from next examples section are visualizations of the hadamard and unitary transform U_f matrices, which are used in each case.

❖ Examples:

```
>> deutsch_jozsa('djc0','htm')
|psi> = 1|000001>
... unitary function matrix
... hadamard transform matrix
... H*Uf*H*|psi>
... measure |psi>
|psi> = 1|000001>
```

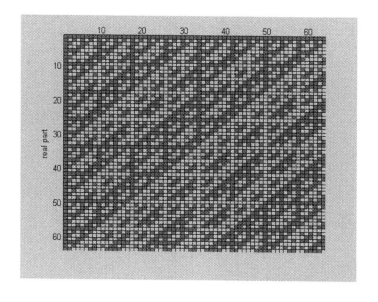

```
>> deutsch_jozsa('djc0','ufm')
|psi> = 1|000001>
... unitary function matrix
... hadamard transform matrix
... H*Uf*H*|psi>
... measure |psi>
|psi> = 1|000000>
```

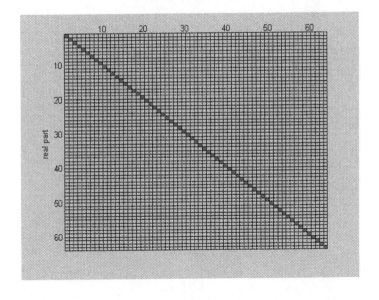

```
>> deutsch_jozsa('djc1','ufm')
|psi> = 1|000001>
... unitary function matrix
... hadamard transform matrix
... H*Uf*H*|psi>
... measure |psi>
|psi> = 1|000001>
```

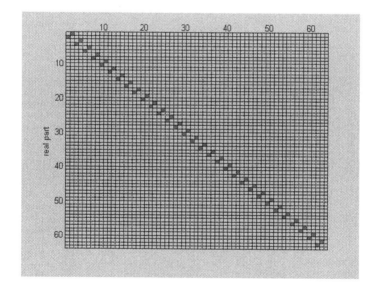

```
>> deutsch_jozsa('djb0','ufm')
|psi> = 1|000001>
... unitary function matrix
... hadamard transform matrix
... H*Uf*H*|psi>
... measure |psi>
|psi> = 1|000011>
```

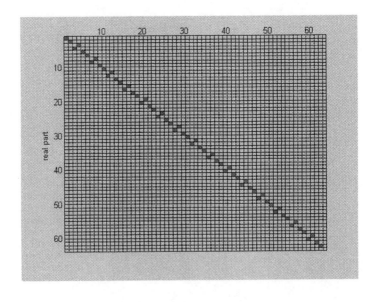

```
>> deutsch_jozsa('djb1','ufm')
|psi> = 1|000001>
... unitary function matrix
... hadamard transform matrix
... H*Uf*H*|psi>
... measure |psi>
|psi> = 1|000011>
```

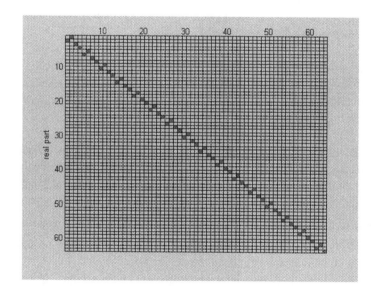

5.4   Grover's Algorithm

Grover's algorithm is a quantum algorithm for searching an unsorted database with n entries in $O(n^{1/2})$ time and using $O(\log n)$ storage space. It was invented by Lov Grover in 1996.

Classically, searching an unsorted database requires a linear search, which is $O(n)$ in time. Grover's algorithm, which takes $O(n^{1/2})$ time, is the fastest possible quantum algorithm for searching an unsorted database. It provides a quadratic speedup, unlike other quantum algorithms, which can provide exponential speedup over their classical counterparts. However, even quadratic speedup is considerable when n is large.

Like all quantum computer algorithms, Grover's algorithm is probabilistic, in the sense that it gives the correct answer with high probability. The probability of failure can be decreased by repeating the algorithm.

Although the purpose of Grover's algorithm is usually described as *searching a database*, it may be more accurate to describe it as *inverting a function*. Roughly speaking, if we have a function $y = f(x)$ that can be evaluated on a quantum computer, Grover's algorithm allows us to calculate x when given y. Inverting a function is related to the searching of a database because we could come a function that produces a particular value of y if x matches a desired entry in a database, and another value of y for other values of x.

This problem, too, can be formulated as an oracle or *black box* problem. In this case, the oracle can be like the phone book, or like a lookup table. We can input a name (a value of x) and the oracle outputs either 0, if $f(x) \neq y$, or 1, if $f(x) = y$. Our task is to find, as quickly as possible, the value of x with $f(x) = y$. Why is this problem important? You may have never tried to find in the phone book the name that matches a given number, but if it were not so hard you might try it more often! More broadly, a rapid method for searching an unsorted database could be invoked to solve any problem in NP.

Our oracle could be a subroutine that interrogates every potential *witness* y that could potentially testify to certify a solution to the problem. For example, if we are confronted by a graph and need to know if it admits a Hamiltonian path, we could submit a path to the *oracle*, and it could quickly answer whether the path is Hamiltonian or not. If we knew a fast way to query the oracle about all the possible paths, we would be able to find a Hamiltonian path efficiently.

So *oracle* could be shorthand for a subroutine that quickly evaluates a function to check a proposed solution to a decision problem, but let us continue to regard the oracle abstractly, as a black box. The oracle *knows* that of the 2^n possible strings of length n, one (the *marked* string or *solution*, ω) is special. We submit a query x to the oracle, and it tells us whether $x = \omega$ or not. It returns, in other words, the value of a function $f_\omega(x)$, with

$$f_\omega(x) = 0, \qquad x \neq \omega$$
$$f_\omega(x) = 1, \qquad x = \omega$$

But furthermore, it is a quantum oracle, so it can respond to queries that are superpositions of strings. The oracle is a quantum black box that implements the unitary transformation

$$U_{f_\omega} : |x\rangle|y\rangle \to |x\rangle|y \oplus f(x)\rangle$$

where $|x\rangle$ is an n-qubit state, and $|y\rangle$ is a single-qubit state.

As we have previously seen in other contexts, we may choose the state of the single-qubit register to be $1/\sqrt{2}\,(|0\rangle - |1\rangle)$, so that the oracle acts as

121

$$U_{f_\omega} : |x\rangle \frac{1}{\sqrt{2}} (|0\rangle - |1\rangle) \rightarrow (-1)^{f_\omega(x)} |x\rangle \frac{1}{\sqrt{2}} (|0\rangle - |1\rangle)$$

We may now ignore the second register, and obtain

$$U_\omega : |x\rangle \rightarrow (-1)^{f_\omega(x)} |x\rangle$$

or

$$U_\omega : 1 - 2|\omega\rangle\langle\omega|$$

The oracle flips the sign of the state $|\omega\rangle$, but acts trivially on any state orthogonal to $|\omega\rangle$. This transformation has a simple geometrical interpretation. Acting on any vector in the $2n$-dimensional Hilbert space (H), U_ω reflects the vector about the hyperplane orthogonal to $|\omega\rangle$ (it preserves the component in the hyperplane, and flips the component along $|\omega\rangle$).

We know that the oracle performs this reflection for some particular computational basis state $|\omega\rangle$, but we know nothing a priori about the value of the string ω. Our job is to determine ω, with high probability, consulting the oracle a minimal number of times.

Consider an unsorted database with n entries. The algorithm requires an n-dimensional state space H, which can be supplied by $\log_2 n$ qubits.

Let us number the database entries by $0, 1, ..., n-1$. Choose an observable, Ω, acting on H, with n distinct eigenvalues whose values are all known. Each of the eigenstates of Ω encode one of the entries in the database, in a manner that we will describe. Denote the eigenstates (using Dirac notation) as

$$\{|0\rangle, |1\rangle, ..., |n-1\rangle\}$$

and the corresponding eigenvalues by

$$\{\lambda_0, \lambda_1, ..., \lambda_{n-1}\}$$

We are provided with a unitary operator, U_ω, which acts as a subroutine that compares database entries according to some search criterion. The algorithm does not specify how this subroutine works, but it must be a *quantum* subroutine that works with superpositions of states. Furthermore, it must act specially on one of the eigenstates, $|\omega\rangle$, which corresponds to the database entry matching the search criterion. To be precise, we require U_ω to have the following effects:

$$U_\omega |\omega\rangle = -|\omega\rangle$$
$$U_\omega |x\rangle = |x\rangle \qquad \text{for all } x \neq \omega$$

Our goal is to identify this eigenstate $|\omega\rangle$, or equivalently the eigenvalue ω, that U_ω acts specially upon.

The steps of Grover's algorithm are as follows:

o Initialize the system to the state

$$|s\rangle = \frac{1}{\sqrt{n}} \sum_x |x\rangle$$

o Perform the following *Grover iteration* $r(n)$ times. The function $r(n)$ is described below.

■ Apply the operator: U_ω

■ Apply the operator: $U_s = 2|s\rangle\langle s| - I$

123

- Perform the measurement Ω. The measurement result will be λ_ω with probability approaching 1 for $n \gg 1$. From λ_ω, ω may be obtained.

To explain the altgorithm recall our initial state

$$|s\rangle = \frac{1}{\sqrt{n}} \sum_x |x\rangle$$

Consider the plane spanned by $|s\rangle$ and $|\omega\rangle$. Let $|\omega^\perp\rangle$ be a ket in this plane perpendicular to $|\omega\rangle$. Since $|\omega\rangle$ is one of the basis vectors, the overlap is

$$\langle \omega | s \rangle = \frac{1}{\sqrt{n}}$$

In geometric terms, there is an angle $(\pi/2 - \theta)$ between $|\omega\rangle$ and $|s\rangle$, where θ is given by:

$$\cos\left(\frac{\pi}{2} - \theta\right) = \frac{1}{\sqrt{n}}, \text{ and } \sin\theta = \frac{1}{\sqrt{n}}$$

The operator U_ω is a reflection at the hyperplane orthogonal to $|\omega\rangle$; for vectors in the plane spanned by $|s\rangle$ and $|\omega\rangle$, it acts as a reflection at the line through $|\omega^\perp\rangle$. The operator U_s is a reflection at the line through $|s\rangle$. Therefore, the state vector remains in the plane spanned by $|s\rangle$ and $|\omega\rangle$ after each application of U_s and after each application of U_ω, and it is straighforward to check that the operator $U_s U_\omega$ of each

Grover iteration step rotates the state vector by an angle of 2θ toward $|\omega\rangle$.

We need to stop when the state vector passes close to $|\omega\rangle$; after this, subsequent iterations rotate the state vector *away* from $|\omega\rangle$, reducing the probability of obtaining the correct answer. The number of times to iterate is given by r. In order to align the state vector exactly with $|\omega\rangle$, we need:

$$\frac{\pi}{2}-\theta=2\theta r, \quad \text{and} \quad r=\frac{\left(\dfrac{\pi}{\theta}-2\right)}{4}$$

However, r must be an integer, so generally we can only set r to be the integer closest to $(\pi/\theta-2)/4$. The angle between $|\omega\rangle$ and the final state vector is $O(\theta)$, so the probability of obtaining the wrong answer is $O(1-\cos^2\theta)=O(\sin^2\theta)$.

For $n \gg 1$, $\theta \approx n^{-1/2}$, so

$$r \to \frac{\pi\sqrt{n}}{4}$$

Furthermore, the probability of obtaining the wrong answer becomes $O(1/n)$, which goes to zero for large n.

If, instead of 1 matching entry, there are k matching entries, the same algorithm works but the number of iterations must be $\pi(n/k)^{1/2}/4$ instead of $\pi n^{1/2}/4$. There are several ways to handle the case if k is unknown. For example, one could run Grover's algorithm several times, with

$$\pi \frac{n^{1/2}}{4}, \pi \frac{(n/2)^{1/2}}{4}, \pi \frac{(n/4)^{1/2}}{4}, ...$$

iterations. For any k, one of iterations will find a matching entry with a sufficiently high probability. The total number of iterations is at most

$$\pi \frac{n^{1/2}}{4} \left(1 + \frac{1}{\sqrt{2}} + \frac{1}{2} + ... \right)$$

which is still $O(n^{1/2})$.

It is known that Grover's algorithm is optimal. That is, any algorithm that accesses the database only by using the operator U_ω must apply U_ω at least as many times as Grover's algorithm.

Below, we present the basic form of Grover's algorithm, which searches for a single matching entry. In this example of Grover's algorithm, we use a function *hay* (~/qct_e/grover/) to represent the database. We are searching for a *needle in the haystack*, i.e. there is one element of the database that we require. The *hay* function returns 1 if queried with the correct *needle* element, and 0 for all the other elements.

hay.m	
1	function h = hay(cond,i)
2	h = 0;
3	if i == cond
4	h=1;
5	end
6	

The listing for Matlab code of the Grover's algorithm
(~/qct_e/grover/) is given below.

```
grover.m

1    function grover(n,p,arg)
2    % grover -- Grover's algorithm
3    %
4    % written by Ioan Burda, (c) 2005
5
6    if ~nargin
7        error('usage: grover(n,[p,psi/iter/htm/iam/uim]')
8    end
9    if n > 8
10       error('qubits number too big');
11   end
12   if (nargin == 1)
13       arg = 'none';
14       p = 1;
15   end
16   if (nargin == 3)
17       h = figure(1);
18       set(h,'MenuBar','none','NumberTitle','off');
19   if strcmp(arg,'psi')
20       set(h,'Name','Grover''s Algorithm -- psi');
21     elseif strcmp(arg,'iter')
22       set(h,'Name','Needle values during algorithm iterations');
23     elseif strcmp(arg,'uim')
24       set(h,'Name','Unitary f-conditional Inverter Matrix');
25     elseif strcmp(arg,'htm')
26       set(h,'Name','Hadamard Transform mMtrix');
27     elseif strcmp(arg,'iam')
28       set(h,'Name','Inversion about Average Matrix');
29   end
30   end
31   %
32   needle = floor((2^n-1)*rand);
33   disp(['needle = ',num2str(needle)]);
34   %
```

```
35    psi = mket(0,n);
36    disp(['|psi> = ', dket(psi)]);
37    %
38    disp('... unitary f-conditional inverter matrix');
39    hay  = uim('hay',needle,n);
40    if strcmp(arg,'uim')
41       igm(hay);
42    end
43    %
44    disp('... hadamard transform matrix');
45    H = htm(n);
46    if strcmp(arg,'htm')
47       igm(H);
48    end
49    %
50    disp('... inversion about average matrix');
51    D = iam(n);
52    if strcmp(arg,'iam')
53       igm(D);
54    end
55    %
56    maxi = fix(p*(pi/4)*sqrt(2^n));
57    disp('... iteration');
58    psi = H*psi;
59    for i=1:maxi
60       psi=hay*psi;
61       psi=D*psi;
62       per(:,i) =  psi;
63       if strcmp(arg,'psi')
64          pket(psi);
65       end
66    end
67    %
68    if strcmp(arg,'iter')
69    surf(per);
70    set(gca,'ZLimMode','manual','ZLim',[-1 1]);
71    set(gca,'YLimMode','manual','YLim',[0 2^n-1]);
72    shading interp;
73    colormap('cool');
74    end
```

75	%
76	disp('... measure |psi>');
77	psi=measure(psi);
78	disp(['|psi> = ',dket(psi,'dec')]);
79	

The images from next examples section are visualizations of the final *psi* state, unitary inverse matrix, inverse about average matrix and amplitude evolution (iter) for five predicted iteration, which are used in each case.

❖ Examples:

```
>> grover(5,5,'psi')
needle = 12
|psi> = 1|00000>
... unitary f-conditional inverter matrix
... hadamard transform matrix
... inversion about average matrix
... iteration
... measure |psi>
|psi> = 1|12>
```

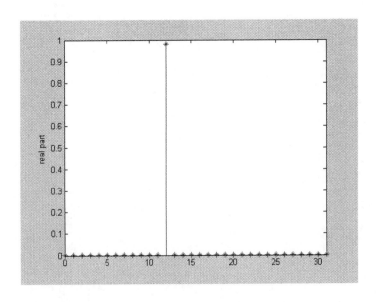

```
>> grover(5,5,'uim')
needle = 28
|psi> = 1|00000>
... unitary f-conditional inverter matrix
... hadamard transform matrix
... inversion about average matrix
... iteration
... measure |psi>
|psi> = 1|28>
```

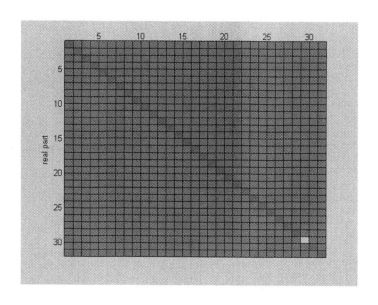

```
>> grover(5,5,'iam')
needle = 27
|psi> = 1|00000>
... unitary f-conditional inverter matrix
... hadamard transform matrix
... inversion about average matrix
... iteration
... measure |psi>
|psi> = 1|27>
```

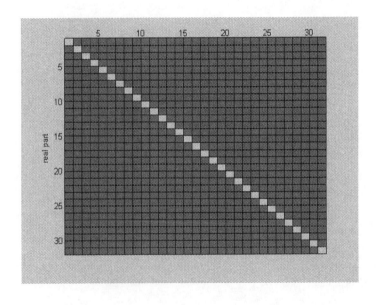

```
>> grover(8,5,'iter')
needle = 59
|psi> = 1|00000000>
... unitary f-conditional inverter matrix
... hadamard transform matrix
... inversion about average matrix
... iteration
... measure |psi>
|psi> = 1|59>
```

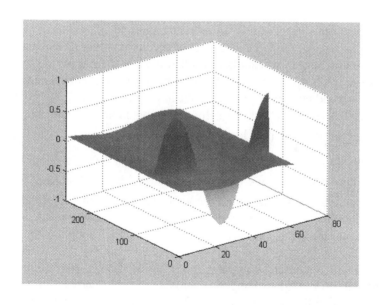

5.5   Shor's Algorithm

By the early nineties, it was know that a quantum computer could be faster than any classical computer. Nonetheless, these observations were largely driven by academic curiosity. There was not much motive for people to spend lots of money or time trying to build a quantum computer.

That changed in 1994 when Peter Shor, a scientist working for Bell Labs devised a polynomial time algorithm for factoring large numbers on a quantum computer. This discovery drew great attention to the field of quantum computing.

The algorithm was viewed as important because the difficulty of factoring large numbers is relied upon for most of the world's cryptography systems. If an efficient method of factoring large numbers were, implemented most of the current encryption scenes would be next to worthless to protect their data from prying eyes. While it has not been proven that factoring large numbers can not be archived on a classical computer in polynomial time, the fastest algorithm publicly available for factoring large number runs in $O(\exp(c(\log n)^{1/3}(\log\log n)^{2/3}))$, or exponential time. In contrast, Shor's algorithm runs in $O((\log n)^2(\log\log n))$ on a quantum computer, and then must perform $O(\log n)$ steps of post processing on a classical computer. Overall then this time is polynomial. This discovery propelled the study of quantum computing forward; as such, an algorithm is much sought after.

Like all quantum computer algorithms, Shor's algorithm is probabilistic: it gives the correct answer with high probability, and the probability of failure can be decreased by repeating the algorithm. Shor's algorithm for factoring a given integer n can be broken into some simple steps.

- o   Determine if the number n is a prime, an even number, or an integer power of a prime number. If it is we will not use Shor's algorithm. There are efficient classical methods for determining if an integer n belongs to one of the above groups, and providing factors for it. This step would be performed on a classical computer.

o Pick an integer q that is the power of 2 such that $n^2 \leq q < 2n^2$. This step would be done on a classical computer.

o Pick a random integer x that is coprime to n. When two numbers are coprime it means that, their greatest common divisor is 1. There are efficient classical methods for picking such an x. This step would be done on a classical computer.

o Create a quantum register and partition it into two sets, register 0 and register 1. Thus the state of our quantum computer can be given by: $|reg1, reg2\rangle$. Register 1 must have enough qubits to represent integers as large as $q-1$. Register 2 must have enough qubits to represent integers as large as $n-1$. The calculations for how many qubits are needed would be done on a classical computer.

o Load register 1 with an equally weighted superposition of all integers from $0 \div q - 1$. Load register 2 with all zeros. Our quantum computer would perform this operation. The total state of the quantum memory register at this point is:

$$\frac{1}{\sqrt{q}} \sum_{a=0}^{q-1} |a, 0 >$$

o Apply the transformation $x^a \bmod n$ to for each number stored in register 1 and store the result in register 2. Due to quantum parallelism this will take only one step, as the quantum computer will only calculate $x^{|a\rangle} \bmod n$, where $|a\rangle$ is the superposition of states created in above step. This step is performed on the quantum computer. The state of the quantum memory register at this point is:

$$\frac{1}{\sqrt{q}} \sum_{a=0}^{q-1} | a, x^a \bmod n >$$

o   Measure register 2, and observe some value k. This has the side effect of collapsing register 1 into a equal superposition of each value a between 0 and $q-1$ such that $x^{|a\rangle} \bmod n = k$. The quantum computer performs this operation. The state of the quantum memory register after this step is:

$$\frac{1}{\sqrt{\|A\|}} \sum_{a'=a'\in A} | a', k >$$

Where A is the set of a's such that $x^{|a\rangle} \bmod n = k$, and $\|A\|$ is the number of elements in that set.

o   Compute the discrete Fourier transform on register 1. The discrete Fourier transform when applied to a state $|a\rangle$ changes it in the following manner:

$$| a > = \frac{1}{\sqrt{q}} \sum_{c=0}^{q-1} | c > * e^{2\pi i a c / q}$$

This step is performed by the quantum computer in one step through quantum parallelism. After the discrete Fourier transform our register is in the state:

$$\frac{1}{\sqrt{\|A\|}} \sum_{a'\in A} \frac{1}{\sqrt{q}} \sum_{c=0}^{q-1} | c, k > * e^{2\pi i a' c / q}$$

o   Measure the state of register one, call this value m, this integer m has a very high probability of being a multiple

136

of q/r, where r is the desired period. The quantum computer performs this step.

o Take the value m, and on a classical computer do some post processing which calculates r based on knowledge of m and q. There are many ways to do this post processing. This post processing is done on a classical computer.

o Once you have attained r, a factor of n can be determined by taking $\gcd(x^{r/2}+1,n)$ and $\gcd(x^{r/2}-1,n)$. If you have found a factor of n, then stop, if not pick a new random integer x and run again. This final step is done on a classical computer.

The algorithm is composed of two parts. The first part of the algorithm turns the factoring problem into the problem of finding the period of a function, and may be implemented classically. The second part finds the period using the quantum Fourier transform, and is responsible for the quantum speedup.

Shor's period-finding algorithm relies heavily on the ability of a quantum computer to be in many states simultaneously. Physicists call this behaviour a *superposition* of states. To compute the period of a function f, we evaluate the function at all points simultaneously.

Quantum physics does not allow us to access all this information directly, though. A measurement will yield only one of all possible values, destroying all others. Therefore we have to carefully transform the superposition to another state that will return the correct answer with high probablity. This is achieved by the quantum Fourier transform.

For positive x, n such that $x < n$, the order of x modulo n is the least positive integer a such that $x^a \bmod n = 1$.

❖ Example:

Suppose $n = 77$ and $x = 10$

a	0	1	2	3	4	5	6	7	8	9	10	11
$x^a \bmod n$	1	10	23	76	67	54	1	10	23	76	67	54

Suppose $n = 15$ and $x = 2$

a	0	1	2	3	4	5	6	7	8	9
$x^a \bmod n$	1	2	4	8	1	2	4	8	1	2

Suppose $n = 15$ and $x = 4$

a	0	1	2	3	4	5	6	7	8	9
$x^a \bmod n$	1	4	1	4	1	4	1	4	1	4

Suppose $n = 15$ and $x = 11$

a	0	1	2	3	4	5	6	7	8	9
$x^a \bmod n$	1	11	1	11	1	11	1	11	1	11

Observe that the period of repetition is one less than a factor of n.

Continuous fraction expansion is defined below:

$$[a_0, a_1, ..., a_m] = a_0 + \cfrac{1}{a_1 + \cfrac{1}{a_2 + \cfrac{1}{\cdots}}}$$

❖ Examples:

$$\frac{13}{64} = \frac{1}{\dfrac{64}{13}} = \frac{1}{4 + \dfrac{12}{13}} = \frac{1}{4 + \dfrac{1}{\dfrac{13}{12}}} = \frac{1}{4 + \dfrac{1}{1 + \dfrac{1}{12}}} \approx \frac{1}{5}$$

We must describe also the modular exponentiation. In Shor's algorithm the evolution of registers state can be written as

$$|a\rangle|y\rangle \rightarrow |a\rangle|x^a \bmod n\rangle$$

Note that:

$$A = a_t 2^{t-1} + a_{t-1} 2^{t-2} + \ldots + a_1 2^0$$
$$x^A = x^{a_t 2^{t-1}} \times x^{a_{t-1} 2^{t-2}} \times \ldots \times x^{a_1 2^0}$$

We can compute x^{2^i} classically. Note that a_i is binary, hence this is really a sequence of conditional multiplies, almost. The tricky bit is mod, but that distributes, i.e.:

$$[(in + \alpha)k] \bmod n = [in \cdot k + \alpha \cdot k] \bmod n = \alpha \cdot k \bmod n$$

Thus, compute:

$$[x^{2^{t-1}} \bmod n]^{a_t} \times [x^{2^{t-2}} \bmod n]^{a_{t-1}} \times \ldots \times [x^{2^0} \bmod n]^{a_0}$$

Note that this is only t conditional modular multiplications.

Below, we present the Matlab code listing of Shor's algorithm(~/qct_e/shor/).

```
1    function shor(n,arg)
2    % Shor's Algorithm for Quantum Factorization
3    %
4    % written by Ioan Burda, (c) 2005
5
6    if ~nargin
7       error('usage: shor(n,[reg1/reg2]')
8    end
9    %
10   if isoks(n)
11      return
12   end
13   %
14   if (nargin == 1)
15      arg = 'none';
16   end
17   if (nargin == 2)
18   h = figure(1);
19   set(h,'MenuBar','none',...
20       'NumberTitle','off','DoubleBuffer','on');
21      if strcmp(arg,'reg1')
22         set(h,'Name','Quantum Fourier Transform');
23      elseif strcmp(arg,'reg2')
24         set(h,'Name','x^a mod n - reg2');
25      end
26   end
27   %
28   % classical part of the Shor's algorithm
29   %
30   p = 0;
31   q = 0;
32   while ~(p*q == n & p ~= 1)
33   disp ('>>> start');
34   xt = true;
35   while (xt)
36   x = floor(rand*(n-3))+2;
37   if gcd(x,n) == 1
38      xt = false;
```

```
39    end
40    end
41    disp(['# x = ',num2str(x)]);
42    %
43    % quantum part of the Shor's algorithm
44    %
45    disp('# quantum');
46    nreg = ceil(log2(n+1));
47    reg1 = mket(0,nreg);
48    reg2 = mket(0,nreg);
49    disp(['|reg1> = ',dket(reg1)]);
50    disp(['|reg2> = ',dket(reg2)]);
51    %
52    disp('... quantum superposition');
53    reg1 = htm(nreg)*(reg1);
54    %
55    disp('... quantum entangled');
56    reg2 = rxamod(reg2,x,n);
57    if strcmp(arg,'reg2')
58       pket(reg2);
59    end
60    %
61    disp('... measure |reg2>');
62    reg2 = measure(reg2);
63    disp(['|reg2> = ',dket(reg2)]);
64    %
65    disp('... quantum Fourier transform');
66    % *** subspace simulation
67    rv = find(reg2>0)-1;
68    tmp_reg1 = reg1;
69    reg1 = zeros(2^nreg,1);
70    reg1(rv:rv:2^nreg) = tmp_reg1(rv:rv:2^nreg);
71    reg1 = nket(reg1);
72    % ***
73    reg1 = qftm(nreg)*reg1;
74    if (strcmp(arg,'reg1'))
75    stem(abs(reg1),'*g');
76    set(gca,'XLimMode','manual','XLim',[0 2^nreg]);
77    set(gca,'YLimMode','manual','YLim',[0 1]);
78    drawnow;
```

```
79    end
80    %
81    disp('... measure |reg1>');
82    reg1 = measure(reg1);
83    m = find(reg1>0)- 1;
84    disp(['|reg1> = ',dket(reg1)]);
85    %
86    % classical part of the Shor's algorithm
87    %
88    if m > 0
89    disp('# continued fraction algorithm');
90    r = cfa(m,length(reg1));
91    %
92    if (r > 1) & (realmax>(x^(r/2)+1))
93    p = gcd(x^(r/2)+1,n);
94    q = gcd(x^(r/2)-1,n);
95    end
96    end
97    end
98    disp('stop <<<');
99    disp([num2str(p),' * ',num2str(q),' = ',num2str(p*q)]);
100   %
101   %
102   % -------------------------------------------------------------
103   function  err = isoks(n)
104   %
105   % test value of n
106
107   err = false;
108   if n < 15
109      disp('15 is the smallest number');
110      err = true;
111      return
112   end
113   if n > 255
114      disp('255 is the biggest number');
115      err = true;
116      return
117   end
118   if mod(n,2) == 0
```

```
119    disp('number must be odd');
120    err = true;
121    return
122  end
123  if ispn(n)
124    disp('prime number');
125    err = true;
126    return
127  end
128  if ispnp(n)
129    disp('prime power');
130    err = true;
131  end
132  % -----------------------------------------------------------
133  function er = ispn(n)
134  %
135  % test if n is a prime number
136
137  er = true;
138  if n <= 1
139    return
140  end
141  for i = 2:floor(sqrt(n))
142    if mod(n,i) == 0
143      er = false;
144      return
145    end
146  end
147  % -----------------------------------------------------------
148  function er = ispnp(n)
149  %
150  % test if n is a prime power
151
152  er = false;
153  tobig = 0;
154  i = 2;
155  f = 0;
156  while(i <= floor(sqrt(n)) & f == 0)
157    if mod(n,i) == 0
158      f = i;
```

```
159        end
160      i = i + 1;
161    end
162    i = 2;
163    while (~tobig & ~er)
164      if f^i == n
165        er = true;
166      end
167      if f^i > n
168        tobig = 1;
169      else
170        i = i + 1;
171      end
172    end
173    % ------------------------------------------------------------
174    function reg = rxamod(reg,x,n)
175    %
176    % register version of the y = x^a mod n
177
178    nreg = log2(length(reg));
179    reg = mket(xamod(x,1,n),nreg);
180    for i = 2:nreg
181      reg = reg + mket(xamod(x,i,n),nreg);
182    end
183    reg = nket(reg);
184    % ------------------------------------------------------------
185    function val = xamod(x,a,n)
186    %
187    % y = x^a mod n
188
189    val = 1;
190    tmp = mod(x,n);
191    ba = dec2bin(a);
192    for i =length(ba):-1:1
193      if (str2num(ba(i))&1)
194        val = mod(val*tmp,n);
195      end
196      tmp = mod(tmp*tmp,n);
197    end
198    % ------------------------------------------------------------
```

```
199   function dd = cfa(p,qmax)
200   %
201   % continued fraction algorithm
202
203   r=1;
204   i=0;
205   while r>0
206     i=i+1;
207     qq(i)=floor(qmax/p);
208     r=qmax-qq(i)*p;
209     qmax=p;
210     p=r;
211   end
212   for i=length(qq):-1:2
213     a=qq(i);
214     b=1;
215     for j=i-1:-1:1
216       c=a;
217       a=qq(j)*a+b;
218       b=c;
219     end
220     dd(i)=a;
221   end
222   dd(1)=qq(1);
223   dd = max(dd);
224
```

The images from next examples section are visualizations of the *reg1* and *reg2*, for few input numbers.

❖ Examples:

```
>> shor(15,'reg1')
....,
>>> start
# x = 7
# quantum
|reg1> = 1|0000>
|reg2> = 1|0000>
... quantum superposition
... quantum entangled
... measure |reg2>
|reg2> = 1|1101>
... quantum Fourier transform
... measure |reg1>
|reg1> = 1|0100>
# continued fraction algorithm
stop <<<
5 * 3 = 15
```

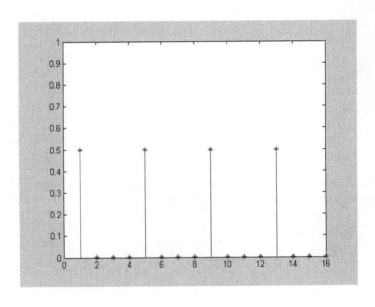

```
>> shor(15,'reg2')
....,
>>> start
# x = 7
# quantum
|reg1> = 1|0000>
|reg2> = 1|0000>
... quantum superposition
... quantum entangled
... measure |reg2>
|reg2> = 1|0100>
... quantum Fourier transform
... measure |reg1>
|reg1> = 1|0100>
# continued fraction algorithm
stop <<<
5 * 3 = 15
```

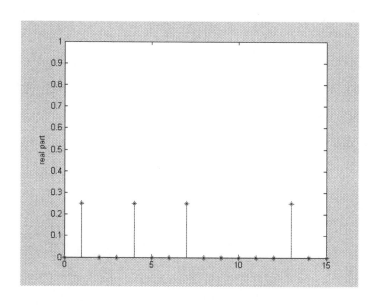

```
>> shor(21,'reg1')
....,
>>> start
# x = 8
# quantum
|reg1> = 1|00000>
|reg2> = 1|00000>
... quantum superposition
... quantum entangled
... measure |reg2>
|reg2> = 1|01000>
... quantum Fourier transform
... measure |reg1>
|reg1> = 1|10000>
# continued fraction algorithm
stop <<<
3 * 7 = 21
```

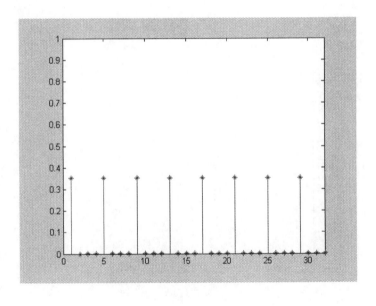

```
>> shor(21,'reg2')
....,
>>> start
# x = 8
# quantum
|reg1> = 1|00000>
|reg2> = 1|00000>
... quantum superposition
... quantum entangled
... measure |reg2>
|reg2> = 1|01000>
... quantum Fourier transform
... measure |reg1>
|reg1> = 1|10000>
# continued fraction algorithm
stop <<<
3 * 7 = 21
```

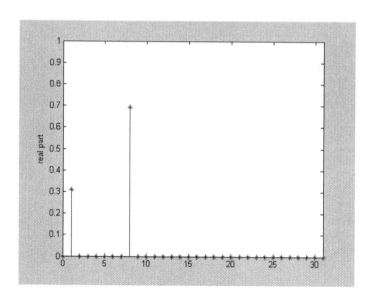

```
>> shor(35,'reg1')
....,
>>> start
# x = 22
# quantum
|reg1> = 1|000000>
|reg2> = 1|000000>
... quantum superposition
... quantum entangled
... measure |reg2>
|reg2> = 1|001000>
... quantum Fourier transform
... measure |reg1>
|reg1> = 1|110000>
# continued fraction algorithm
stop <<<
5 * 7 = 35
```

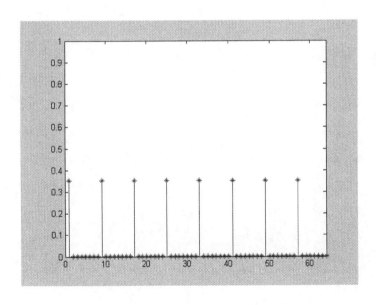

```
>> shor(35,'reg2')
....,
>>> start
# x = 22
# quantum
|reg1> = 1|000000>
|reg2> = 1|000000>
... quantum superposition
... quantum entangled
... measure |reg2>
|reg2> = 1|001000>
... quantum Fourier transform
... measure |reg1>
|reg1> = 1|010000>
# continued fraction algorithm
stop <<<
5 * 7 = 35
```

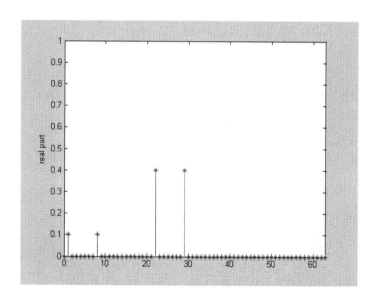

```
>> shor(85,'reg1')
....,
>>> start
# x = 4
# quantum
|reg1> = 1|0000000>
|reg2> = 1|0000000>
... quantum superposition
... quantum entangled
... measure |reg2>
|reg2> = 1|0000100>
... quantum Fourier transform
... measure |reg1>
|reg1> = 1|1100000>
# continued fraction algorithm
stop <<<
17 * 5 = 85
```

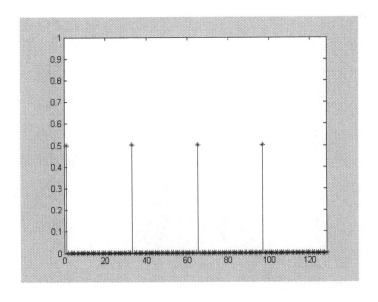

```
>> shor(85,'reg2')
....,
>>> start
x = 4
# quantum
|reg1> = 1|0000000>
|reg2> = 1|0000000>
... quantum superposition
... quantum entangled
... measure |reg2>
|reg2> = 1|1000000>
... quantum Fourier transform
... measure |reg1>
|reg1> = 1|0100000>
# continued fraction algorithm
stop <<<
17 * 5 = 85
```

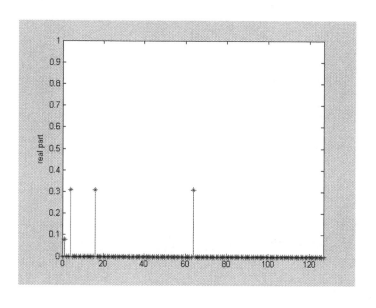

```
>> shor(111,'reg1')
....,
>>> start
# x = 68
# quantum
|reg1> = 1|0000000>
|reg2> = 1|0000000>
... quantum superposition
... quantum entangled
... measure |reg2>
|reg2> = 1|1001001>
... quantum Fourier transform
... measure |reg1>
|reg1> = 1|1100000>
# continued fraction algorithm
stop <<<
37 * 3 = 111
```

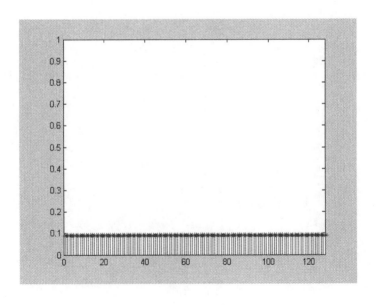

```
>> shor(111,'reg2')
....,
>>> start
# x = 68
# quantum
|reg1> = 1|0000000>
|reg2> = 1|0000000>
... quantum superposition
... quantum entangled
... measure |reg2>
|reg2> = 1|1000100>
... quantum Fourier transform
... measure |reg1>
|reg1> = 1|0100000>
# continued fraction algorithm
stop <<<
37 * 3 = 111
```

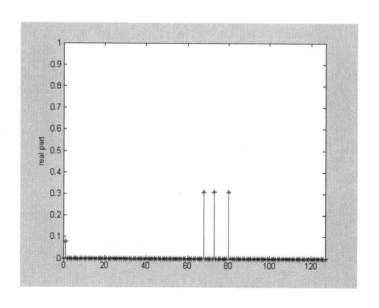

```
>> shor(255,'reg1')
....,
>>> start
# x = 7
# quantum
|reg1> = 1|00000000>
|reg2> = 1|00000000>
... quantum superposition
... quantum entangled
... measure |reg2>
|reg2> = 1|00010000>
... quantum Fourier transform
... measure |reg1>
|reg1> = 1|01110000>
# continued fraction algorithm
stop <<<
17 * 15 = 255
```

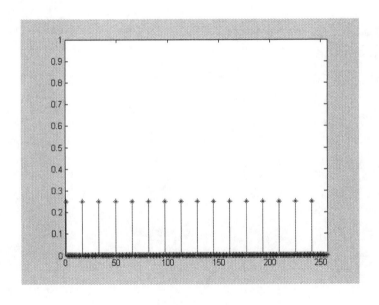

```
>> shor(255,'reg2')
....,
>>> start
# x = 7
# quantum
|reg1> = 1|00000000>
|reg2> = 1|00000000>
... quantum superposition
... quantum entangled
... measure |reg2>
|reg2> = 1|00010000>
... quantum Fourier transform
... measure |reg1>
|reg1> = 1|00010000>
# continued fraction algorithm
stop <<<
17 * 15 = 255
```

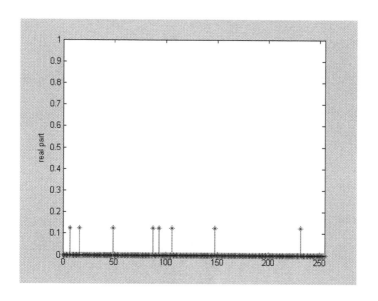

Bibliography

Turing, A.M. (1936). *"On Computable Numbers, With an Application to the Entscheidungsproblem"*, Proceedings of the London Mathematical Society, (2) 42, pp 230-265.

Church, A. (1937). *"Review of Turing 1936.* Journal of Symbolic Logic", 2, pp 42-43.

Deutsch, David (1985). *"Quantum theory, the Church-Turing principle and the universal quantum computer"*, Proceedings of the Royal Society London, A 400, pp 97-117.

Scott, N.R. (1985). *"Computer Number Systems and Aritmetic"*, Englewood Cliffs, NJ: Printice Hall.

Peterson, W.W. and Welden, E.J. (1972). *"Error-Correcting Codes"*, 2nd ed. Boston: MIT Press.

DeMicheli, G. (1994). *"Synthesis and Optimization of Digital Circuits"*, New York: McGraw-Hill.

Kohavi, Z. (1998). *"Switching and Autamotion Theory"*, 2nd ed. New York: McGraw-Hill.

Nielsen, Michael, Chuang A., Isaac, L. (2000). *"Quantum Computation and Quantum Information"*, Cambridge University Press.

Paul, Benio, (1997). *"Models of Quantum Turing Machines"*, LANL Archive quant-ph/9708054.

Bell, J.S. (1964). *"On the Einstein-Podolsky-Rosen paradox"*, Physics 1, 195–200.

Cirac, J.I., Zoller, P. (1995). *"Quantum Computations with Cold trapped Ions"*, Phys. Rev. Lett. 74, 4091.

Deutsch, David (1989). *"Quantum computational networks"*, Proceedings of the Royal Society London A 439, 553-558.

Coppersmith, D. (1994). *"An Approximate Fourier Transform Useful in Quantum Factoring"*, IBM Research Report No. RC19642.

Bennet, C.H. (1973). *"Logical Reversibility of Computation"*, IBM J. Res.Develop. 17, 525.

Murnaghan, F.D. (1962). *"The Unitary and Rotation Groups"*, Spartan Books, Washington.

Ekert, Artur and Jozsa, Richard (1996). *"Shor's Quantum Algorithm for Factoring Numbers"*, Rev. Mod. Physics 68 (3), 733-753.

Hardy, G.H. and Wright, E.M. (1965). *"An Introduction to the Theory of Numbers"*, (4th edition OUP).

Copeland, B.J. (1996). *"The Church-Turing Thesis. Stanford Encyclopedia of Philosophy"*, ISSN 1095-5054.

Turing, A.M. (1948). *"Intelligent Machinery"*, National Physical Laboratory Report, (1948) In Meltzer, B., Michie, D. (eds) 1969, Machine Intelligence 5, Edinburgh: Edinburgh University Press, 7.

Boyer, M., Brassard, G., Hoyer, P., Tapp, A. (1996). *"Tight bounds on quantum searching"*, Proceedings PhysComp96.

Ömer, Bernhard (1998). *"A Procedural Formalism for Quantum Computing, master-thesis"*, Technical University of Vienna.

Benioff, P. (1980). *"The Computer as a Physical System: A Microscopic Quantum Mechanical Hamiltonian Model of Computers as Represented by Turing Machines"*, Journal of Statistical Physics, Vol. 22, pp. 563-591.

Berthiaume, Andre and Brassard, Gilles (1992). *"The quantum Challenge to Complexity Theory"*, Proceedings of the 7th IEEE Conference on Structure in Complexity Theory, pp. 132-137.

Brassard, Gilles (1997). *"Searching a Quantum Phone Book"*, Science 1.

Cormen, Thomas, H., Leiserson, Charles, E. and Rivest Ronald L. (1994). *"Introduction to Algorithms"*, St. Louis: McGraw-Hill.

Deutsch, David and Jozsa, Richard (1992). *"Rapid Solution of Problems by Quantum Computation"*, Proceedings Royal Society London, Vol. 439A, pp. 553-558.

Feynman, Richard (1982). *"Simulating Physics with Computers"*, Optics News Vol. 11, pp. 467-488.

Kitaev, A. Yu. (1995). *"Quantum Measurements and the Abelian Stabilizer Problem"*, quant-ph/9511026

OpenQubit (1998). OpenSource, http://www.openqubit.org/

Tucci, R. R. (1998). Qubiter, http://www.ar-tiste.com/qubiter.html

Baker, Greg (1997). Q-gol, http://www.apt.net.au/~greg/q-gol

Grover L.K. (1996). *"A Fast Quantum Mechanical Algorithm for Database Search"*, Proceedings of the 28'th Annual ACM Symposium on the Theory of Computing, pp. 212-219.

Steane, Andrew (1998). *"Quantum Computing"*, Reports on Progress in Physics, vol 61, pp 117-173.

Williams, Colin, P. and Clearwater, Scott, H. (1998). *"Explorations in Quantum Computing"*, New York: Springer-Verlag.

Shor, P.W. (1994). *"Algorithms for quantum computation: Discrete logarithms and factoring"*, Proceedings of the 35th

Annual Symposium on Foundations of Computer Science, IEEE
Computer Society Press.

Deutsch, David, Ekert, Artur (1998). *"Quantum Computation"*,
Physics World, March-1998.

Gruska, J. (1999). *"Quantum Computing"*, McGraw Hill.

Kitaev, A.Y., Shen, A.H. and Vyalyi, M.N. (2002). *"Classical and
Quantum Computation"*, AMS-2002.

Hogg, Tad (1996). *"Quantum Computing and Phase Transitions
in Combinatorial Search"*, Journal of Artificial Intelligence
Research, Vol. 4, pp91-128.

Bernstein, E., and Vazirani, U. (1993). *"Quantum
Complexity Theory"*, Proceedings of the 25th Annual ACM
Symposium on Theory of Computation, ACM, New York,
pp11-20.

Index

CPSIA information can be obtained
at www.ICGtesting.com
Printed in the USA
LVHW040552290620
659255LV00001B/223